T0328256

Model Ecosystems in
Extreme Environments

Astrobiology Exploring Life on Earth and Beyond

Model Ecosystems in Extreme Environments

Series Editors

Pabulo Rampelotto

Richard Gordon

Joseph Seckbach

Volume Editors

Joseph Seckbach

Pabulo Rampelotto

ELSEVIER

ACADEMIC PRESS

An imprint of Elsevier

Academic Press is an imprint of Elsevier
125 London Wall, London EC2Y 5AS, United Kingdom
525 B Street, Suite 1650, San Diego, CA 92101, United States
50 Hampshire Street, 5th Floor, Cambridge, MA 02139, United States
The Boulevard, Langford Lane, Kidlington, Oxford OX5 1GB, United Kingdom

Notices

Knowledge and best practice in this field are constantly changing. As new research and experience broaden our understanding, changes in research methods, professional practices, or medical treatment may become necessary.

Practitioners and researchers must always rely on their own experience and knowledge in evaluating and using any information, methods, compounds, or experiments described herein. In using such information or methods they should be mindful of their own safety and the safety of others, including parties for whom they have a professional responsibility.

To the fullest extent of the law, neither the Publisher nor the authors, contributors, or editors, assume any liability for any injury and/or damage to persons or property as a matter of products liability, negligence or otherwise, or from any use or operation of any methods, products, instructions, or ideas contained in the material herein.

Library of Congress Cataloging-in-Publication Data
A catalog record for this book is available from the Library of Congress

British Library Cataloguing-in-Publication Data
A catalogue record for this book is available from the British Library

ISBN 978-0-12-812742-1

For information on all Academic Press publications
visit our website at https://www.elsevier.com/books-and-journals

Publisher: Candice Janco
Acquisition Editor: Marisa LaFleur
Editorial Project Manager: Hilary Carr
Production Project Manager: Omer Mukthar
Cover Designer: Greg Harris

Typeset by SPi Global, India

Working together
to grow libraries in
developing countries

www.elsevier.com • www.bookaid.org

Contents

Contributors

Corien Bakermans
The Pennsylvania State University, Altoona, PA, United States

Bonnie K. Baxter
Department of Biology, Great Salt Lake Institute, Westminster College, Salt Lake City, UT, United States

J. Baxter
Centre for Microbial Ecology and Genomics, Department of Biochemistry, Genetics and Microbiology, University of Pretoria, Pretoria, South Africa

Nirit Bernstein
Institute of Soil, Water and Environmental Sciences, ARO, Volcani Center, Rishon LeZion, Israel

D.A. Cowan
Centre for Microbial Ecology and Genomics, Department of Biochemistry, Genetics and Microbiology, University of Pretoria, Pretoria, South Africa

Jocelyne DiRuggiero
Department of Biology, The Johns Hopkins University, Baltimore, MD, United States

Lisa A. Emili
The Pennsylvania State University, Altoona, PA, United States

Kenji Ikehara
G&L Kyosei Institute, Nara; The International Institute for Advanced Studies of Japan, Kyoto, Japan

Scott T. Kelley
Department of Biology, San Diego State University, San Diego, CA, United States

C. León-Sobrino
Centre for Microbial Ecology and Genomics, Department of Biochemistry, Genetics and Microbiology, University of Pretoria, Pretoria, South Africa

G. Maggs-Kölling
Gobabeb Research and Training Centre, Walvis Bay, Namibia

L. Martínez-Alvarez
Centre for Microbial Ecology and Genomics, Department of Biochemistry, Genetics and Microbiology, University of Pretoria, Pretoria, South Africa

Victoria Meslier
Department of Biology, The Johns Hopkins University, Baltimore, MD, United States

Aharon Oren
Department of Plant and Environmental Sciences, The Hebrew University of Jerusalem, The Edmond J. Safra Campus, Jerusalem, Israel

J.-B. Ramond
Centre for Microbial Ecology and Genomics, Department of Biochemistry, Genetics and Microbiology, University of Pretoria, Pretoria, South Africa

D.A. Read
Centre for Microbial Ecology and Genomics, Department of Biochemistry, Genetics and Microbiology, University of Pretoria, Pretoria, South Africa

Helga Stan-Lotter
Department of Microbiology, Division of Molecular Biology, University of Salzburg, Salzburg, Austria

A.J. van der Walt
Centre for Microbial Ecology and Genomics, Department of Biochemistry, Genetics and Microbiology, University of Pretoria, Pretoria, South Africa

Richard L. Weiss Bizzoco
Department of Biology, San Diego State University, San Diego, CA, United States

Polona Zalar
Department of Biology, University of Ljubljana, Ljubljana, Slovenia

Who is who in "model ecosystems in extreme environments"

In this small volume, we present ecological systems of extremophilic environments. Our contributors are from more than a half dozen countries. After the publishers' thorough review, a selection of outstanding chapters was chosen for this book.

Extremophiles is now a branch in biology that has developed recently and describes habitats and their organisms that tolerate and thrive in very harsh conditions of life. Astrobiology also addresses life in severe conditions. Included among these extreme dwelling organisms are microorganisms, such as bacteria, cyanobacteria, various algae, plants (see chapter by Bernstein), and animals (such as the Tardigrades).

The life habitats of such harsh conditions are acidic (low pH levels), alkaline (high pH scale), cryophilic, thermophiles, or barophilic (as those organisms living under high hydrostatic pressure as in the bottom of the oceans). Those that live at multiple extreme conditions are called polyextremophiles.

The science of extremophiles related to *Astrobiology* (earlier called exobiology, space biology, Cosmo-biology) is a relativity new science which includes several branches at various temperature levels, halophile conditions, pH effect on organisms, anaerobic conditions, pressure effects, anaerobic conditions, and so on. Recently NASA announced that Mars contains a body of salty water under its ice caps. In addition, we read that some satellites, like Europa—the "ocean moon" of Jupiter, and Enceladus, satellite of Saturn, in our Solar System contain oceans of salty water under their heavy layers of frozen ice. Water plumes are seen in space photos exiting from the ocean of Europa and from other Satellites. There is still a debate whether these subicy oceans could support habitats of microbial life. So far there is no strong evidence of extraterrestrial life. So far, it is only Mother Earth that is known to harbor life.

On Earth, there are a number of salty areas with living microbes inhabiting them. Among the halo-environments are places such as the Dead Sea (Israel), Rio Tinto (Spain), Yellowstone National Park (USA), Atacama Desert (Chile), and McMurdo dry valley (Antarctica). In most of these areas one can find microbial life, Halophiles, living in extreme conditions.

The conclusion is that salty areas have a strong connection to Astrobiology and to the Origin of Life. The hope is that some planets and satellites will be found to contain in their subicy ocean active microorganisms.

Found in several places around the Earth are the acidic thermophilic red algae *Cyanidium caldarium* (and it cohorts; *Galdieria* and *Cyanidioschyzon*); they are distributed ubiquitously at pH 0–4 and at a high temperature up to 57°C. They are considered among the extremophiles.

The extreme habitats described in this volume are hyperhalophiles and hypersaline systems (Oren, Baxter, Bernstein); Arctic with low temperature (Bakermans), Thermophiles (Bizzoco); Desert environments (Cowan, [Ramond]; Endolithic-subsurface microbes (DiRuggiero); model for Astrobiology (Stan-Lotter). This short volume points out that organisms can be found in both extremophilic severe conditions and perhaps as space biology. We thank all the authors and the reviewers for their excellent cooperation.

Joseph Seckbach
Efrat, Israel

References

Seckbach, J., 1994–2015. Series of Cellular Origin, Life in Extreme Habitats and Astrobiology.

Terrestrial systems of the Arctic as a model for growth and survival at low temperatures

1

Corien Bakermans, Lisa A. Emili

The Pennsylvania State University, Altoona, PA, United States

Chapter outline

1 Introduction

With 23 million km^2 of land that is influenced by the presence of permafrost—ground that has remained frozen for more than 2 years (Zhang et al., 2003)—the Arctic allows the examination of growth and survival of soil microorganisms at low temperatures under a variety of conditions. While mean annual temperatures in the Arctic are typically $-5°C$ or less, temperatures can be stable or can fluctuate drastically on a daily and seasonal basis. Long-term exposure to low temperatures occurs in permafrost and, at depth, temperatures are stable at 0 to $-10°C$ in Alaskan permafrost (Brown and

Model Ecosystems in Extreme Environments. https://doi.org/10.1016/B978-0-12-812742-1.00001-5

1

Romanovsky, 2008) and at 0 to $-13°C$ in Siberian permafrost (Vorobyova et al., 1997). Viable microorganisms have been isolated from permafrost (Gilichinsky and Wagener, 1995; Johnson et al., 2007; Vishnivetskaya et al., 2006), demonstrating that survival of these long-term low temperature conditions is frequently realized. Many Arctic soils that are not perennially frozen experience highly variable temperatures, extreme temperature gradients, and regular freezes and thaws. Frequent freeze–thaw results in significant remodulating of soil structure through cryoturbation, frost heave, and other processes. Temperatures in the upper layer of soil (known as the active layer if underlain by permafrost) range from $-50°C$ in January to $+10°C$ in July. Warm summer temperatures cause Arctic soils to thaw anywhere from 20 to 150 cm in depth and stimulate microbial metabolism. And the microorganisms present in these thermally dynamic environments must tolerate the frequent freeze-thaw.

Arctic biomes are relatively young when compared to other biomes—cold temperatures developed at the end of the Pliocene (about 2.6 million years ago) with the onset of glacial conditions. Subsequently, large amounts of permafrost formed in the Arctic particularly during the middle and upper Pleistocene. For example, the oldest ice within permafrost in North America dates to $\sim$740,000 years ago (Froese et al., 2008); while 2–3 million year old permafrost exists in northeastern Siberia (Vishnivetskaya et al., 2000, 2006). The relatively young ages of these Arctic habitats combined with the low temperatures (which slows metabolism) may constrain the evolutionary adaptation of microbial inhabitants to the low temperature conditions.

While united by low temperatures, the heterogeneity of abiotic factors results in a high diversity of terrestrial habitats in the Arctic. For example, Arctic soils make up 75% of soils as defined by the World Reference Base and vary in mineral composition, snow cover, water content, connectivity, type of vegetation, and organic content (Blaud et al., 2015). In this chapter, we examine how the abiotic factors of Arctic soils affect the microbial communities that grow and survive there. A brief overview of adaptations to low temperatures found in isolates from the Arctic is followed by examination of the biogeochemistry and microbial communities of Arctic soils (not including permafrost) for an integrated view into the interactions between environment and inhabitants. Subsequently, we pay special attention to how the additional constraints of permafrost influence microorganisms and briefly review the potential impacts of warming in this sensitive ecosystem. Finally, we draw conclusions about the adaptations and processes best examined and modeled by microbial communities in terrestrial systems of the Arctic. This chapter is one of very few reviews to bring together the perspectives of both soil characteristics and microbiology on the biogeochemistry and ecosystem function of Arctic soils.

2 Adaptation to low temperatures

The adaptations of organisms to low temperatures have been reviewed extensively (Bakermans, 2012; De Maayer et al., 2014; Margesin, 2017) and will only be summarized here. As temperatures decrease, the thermal motion of atoms and molecules decreases causing all processes to slow down; presenting many challenges to the survival and reproduction of microorganisms. At temperatures below the freezing point of water, the amount of liquid water is reduced as ice begins to form. Without liquid water there is no solvent for the biochemistry that makes life possible; for a review of the known, and potential for, microbial metabolic activity in brines, thin films, and other sources of water, see (Stevenson et al., 2015). As temperature decreases, reaction rates decrease while the rigidity of proteins increases. To combat these effects, the molecular structure of proteins from psychrophiles is altered by decreasing stabilizing interactions to increase the disorder of the molecule thereby maintaining function (Feller, 2013). For a detailed review of the general properties, activity, stability, thermodynamic stability, folding, and engineering of cold-adapted proteins, see Gerday (2013). In addition, cell membranes will undergo a phase change from a liquid to a gel at low temperatures. As a gel, membranes are too rigid to allow protein movement and hence function. To maintain fluidity at lower temperatures, most cells increase the amount of unsaturated fatty acids while some bacteria can also increase the amount of methyl branched fatty acids or shorter fatty acid chains (Gerday, 2011; Russell, 2007, 2008). In many psychrophiles, additional proteins are present that assist DNA replication, transcription, and translation; alter membranes and cell walls; combat oxidative stress; and support osmoregulation at low temperatures. Many psychrophilic organisms have a high halotolerance in part due to commonalities between responses to salt stress and cold stress (Gallardo et al., 2016; Schmid et al., 2009; Srimathi et al., 2007; Welsh, 2000).

Most of the aforementioned cold adaptations are found in microorganisms isolated from the Arctic. However, because active layer survival probably exerts a significant selective force on microorganisms, most isolates from Arctic terrestrial environments are eurypsychrophiles that tolerate a broad range of cold temperatures rather than stenopsychrophiles (e.g., Bakermans et al., 2006; Finster et al., 2009; Mykytczuk et al., 2012; Niederberger et al., 2009; Shcherbakova et al., 2013; Suetin et al., 2009). In addition, some terrestrial psychrophiles have multiple copies of genes (isozymes) with different levels of cold adaptation which likely allows the maintenance of enzymatic function over a broader temperature range (Bergholz et al., 2009; Goordial et al., 2016b). Furthermore, resilience to freeze-thaw is common (Mannisto et al., 2009; Wagner et al., 2013). In part due to a limited number of thoroughly characterized isolates and genome sequences and the complexity of metagenomics and metatranscriptomic data from these diverse communities, it is not yet known which characteristics of terrestrial Arctic eurypsychrophiles (e.g., isozymes) are common in soil microbes and which are unique to Arctic inhabitants.

3 Biogeochemical processes in Arctic soils

Examination of the biogeochemistry of Arctic soils provides an integrated view into the interactions between environment and inhabitants and can reveal insights about productivity, nutrient cycling, and weathering; limitations and temporal variation thereof; and how these processes contribute to ecosystem function and regulation of global climate. Here, the term Arctic soil is used to refer to soils that are not perennially frozen; permafrost is specifically examined in Section 4.

3.1 Trends

The biogeochemistry of terrestrial ecosystems in the Arctic is complex and varies spatially across the landscape (Herndon et al., 2015; Kelley et al., 2012; Whittinghill and Hobbie, 2012), vertically within the soil profile (Kim et al., 2016; Miller et al., 2015; Siewert et al., 2016), and temporally with seasonal hydro-meteoric shifts (Christiansen et al., 2017; Edwards and Jefferies, 2013). At the landscape scale, processes such as nutrient cycling and net primary productivity are driven by macroclimatic factors and vary across a gradient from north to south that is divided into three regions: the High Arctic, the Low Arctic, and the Subarctic. This north to south gradient is characterized by increasing temperature and precipitation (Przybylak, 2000) with concomitant shifts in vegetation communities from dominantly nonvascular to a diverse mix of nonvascular and vascular species (Walker, 2000). Therefore the High Arctic is characterized by deserts, while the tundra of the Low Arctic and Subarctic can be separated by the tree line. All regions contain large areas of permafrost that become more isolated and sporadic as temperatures increase.

Superimposed on this landscape scale north-south gradient is a mosaic of soil environments and geomorphic features formed in response to:

- variations in microtopography over short distances (Iturrate-Garcia et al., 2016; Kelley et al., 2012) and
- the structural heterogeneity of permafrost (both inter and intra-site variation) that contributes to variation in water drainage, decomposition rates, and organic matter accumulation (Bockheim, 2015; Jansson and Tas, 2014).

Consequently, site-specific moisture and biogeochemical regimes develop. In low relief environments, water saturation leads to anoxic conditions that limit decomposition, contributing to the accumulation of organic material and the release of carbon as carbon dioxide (CO_2) and methane (CH_4) through anaerobic microbial metabolism (Lipson et al., 2010). At elevated sites, well-drained soils remain oxic allowing aerobic respiration to facilitate the decomposition of soil organic matter (SOM) with carbon released primarily as CO_2, which limits the thickness of the organic soil horizon (Sturtevant and Oechel, 2013).

Similarly, during seasonal thaw conditions in permafrost-affected soils, the extent of the oxic zone varies with the position of the water table in the active layer. Anoxic conditions prevail in the active layer of low lying sites with high ice content (Mackelprang et al., 2016) and increasingly oxic conditions occur due to lowering of the water table in more well-drained sites (Herndon et al., 2015). Distinct vertical gradients in soil redox potential (Eh) can develop as the water table lowers increasing the depth of the oxidation front, or as the water table rises increasing reducing conditions (Husson, 2013). There is a limited understanding of seasonally controlled redox dynamics due to the spatial variation in onset and rates of thaw (Edwards and Jefferies, 2013), the lack of a distinct line of transition from oxic to anoxic conditions due to variation in soil physical properties (Jorgensen et al., 2015), and difficulties in measuring Eh in the field (Street et al., 2016). Mackelprang and others (2016) suggest that microbial data may best directly indicate Eh microsite heterogeneity. In Arctic soils, iron is often the most prevalent redox sensitive species with alternative electron acceptors, for example, nitrate, sulfate, and manganese being very low in concentration. An extensive analysis of the interaction of microbial-mediated iron cycling in tundra soils indicated that iron reduction dominates anaerobic respiration in organic soils with abundant iron oxides or organically bound Fe(III); potentially contributing 43%–63% of ecosystem respiration (Lipson et al., 2010, 2013, 2015). Methanogenesis may also be limited as Fe-reducing bacteria that produce CO_2 out-compete methanogens for carbon substrates (Lipson et al., 2012). A negative mechanistic link between Fe reduction and CH_4 production was also found in wetland areas of tundra (Miller et al., 2015).

Across the mosaic of site-specific conditions of Arctic soils, there is a distinct vertical stratification in pH and solute chemistry, whereby pH and inorganic nutrient concentration (NO_3^-, PO_4^{3-}, Ca^{2+}, Mg^{2+}, K^+) increase with depth from the organic horizon to the mineral horizon (McCann et al. 2016) and in permafrost compared to the active layer of mineral soils (Bockheim et al., 1998; Keller et al. 2007). Nutrient cycling is intimately tied to the accumulation of organic matter in the organic horizon and to weathering processes in the mineral horizon. The organic horizon is characterized by a high level of acidity due to incomplete decomposition and the production of humic and fulvic acids with high cation exchange capacities (75%–90% of total capacity) that effectively remove nutrient base cations (Ca^{2+}, Mg^{2+}, Na^+, K^+) from soil-water suspensions (Prescott et al., 1995), limiting the availability of these nutrients for plant growth. The base cation concentration of organic soils can also vary spatially across the landscape due to deposition of unweathered sediment by fluvial, aeolian (loess), or glacial processes (Keller et al. 2007; Whittinghill and Hobbie, 2012). Over time these cations are removed from the organic horizon by near surface weathering processes, cation uptake, and leaching to the subsoil mineral horizon. The active layer of mineral soils can also be depleted in cations by progressive weathering in that layer (Stutter and Billett, 2003; Whittinghill and Hobbie, 2012).

Organic matter may be present in the mineral horizon as a consequence of cryogenic processes such as freeze-thaw and in permafrost as a relic from a previous climatic environment with higher plant productivity (Siewert et al., 2016). In the case of mineral soils with organics, mineral concentration is a function of weathering processes and/or the input of ion-rich water from depth during periodic thawing of the upper permafrost (Kokelj and Burn, 2005). In the permafrost, cycling is likely suspended under frozen conditions (Anisimov et al., 1997; and see Section 5). However, warming soil conditions, increased active layer thickness, and deeper thaw may increase the upward movement of base cations by cryoturbation and melting of permafrost causing an increase in pH, changes to microbial substrate availability (Miller et al., 2015; Whittinghill and Hobbie, 2012), and changes in vegetation community composition (Iturrate-Garcia et al., 2016; Keller et al., 2007). These changes may be mitigated by increased organic matter decomposition and organic acid production (Davidson and Janssens, 2006).

3.2 Nutrient limitation

Terrestrial Arctic ecosystems are nutrient limited either by phosphorus alone or by both phosphorus (P) and nitrogen (N). Both P and N vary spatially across the landscape as a function of organic matter accumulation and moisture conditions. Wet, acidic sites tend toward colimitation (e.g., Nadelhoffer et al., 2002) and more mesic, circumneutral sites tend toward N limitation (Turner et al., 2004). Organic nutrient pools are a function of litter quality (Vincent et al., 2014) and external inputs, that is, deposition, weathering and microbial fixation, although the input rates are lower than internal mineralization rates (Shaver et al., 1992). Tundra soils contain large pools of nutrients; however, these nutrients are immobilized in recalcitrant SOM, plant tissue with slow turnover rates, and microbial biomass (Pearce et al., 2015; Wu et al., 2006).

Nitrogen cycling in Arctic soils is relatively well understood and similar to other soils. Nitrogen enters the soil through symbiotic and free-living microbial N-fixation (Rousk et al., 2016). Due to the high C:N ratio of undecomposed organic matter in organic soil horizons, there is a net N immobilization by microbial assimilation (Hobbie and Gough, 2002). In well-drained, oxic sites, mineralized nitrogen (ammonium) is available for nitrification during seasonal thaws (Reyes and Lougheed, 2015). The efflux of nitrates from the soil occurs by denitrification or leaching (Edwards and Jefferies, 2013). Under elevated temperature and moisture conditions higher N availability due to enhanced N-fixation may further increase mineralization, nitrification rates, and available NO_3^- (Penton et al., 2016). Higher N fixation might also increase denitrification rates if water tables rise, leading to an increase in gas effluxes (Penton et al., 2016).

Phosphorus cycling in Arctic soils is less well understood (Turner et al., 2004). Organic P is the predominant source of plant and microbial P. Nearly one-third of the total P in Arctic soils is contained in microbial biomass (Jonasson et al., 1996) and stresses related to wetting/drying or freeze-thaw cycles could significantly impact P availability by promoting the release of P (Buckeridge and Grogan, 2010). The amount of available inorganic P (phosphate) represents a small pool, increasing under warmer, more oxic conditions that promote mineralization. As saturation increases in the organic horizon, phosphate (PO_4^{3-}) concentration decreases due to the competing processes of plant uptake and adsorption by organic acids (Prescott et al., 1995). Permafrost contains a comparatively greater concentration of inorganic P in comparison to the active layer and presents a large pool of potentially available P (Gray et al., 2014).

3.3 Microbial communities

Overall, microbial communities from Arctic soils are similar to microbial communities from other soils, for a recent review, see Blaud et al. (2015). For example, in a survey of 29 heath tundra sites at least 100 km apart across the Arctic, microbial diversity was found to be as high as temperate soils (Chu et al., 2010). The dominate microbial phyla at these heath tundra sites included Acidobacteria, Alphaproteobacteria, Actinobacteria, Betaproteobacteria, and Bacteroidetes; community composition correlated most with pH and C:N ratio (Chu et al., 2010). Microbial communities do not vary significantly by latitude; indeed, the diversity of microbial communities in soils has been found to vary primarily with pH rather than latitude or other factors (Fierer and Jackson, 2006). However, because there are currently no standard protocols for the analysis of microbial communities in soils (in part due to rapidly changing technologies), our ability to interpret microbial community data, make comparisons between studies, and draw conclusions is limited.

Not surprisingly, some of the trends seen in the biogeochemistry of Arctic soils (a mosaic of site-specific conditions and vertical layers) are also found in the microbial communities that are often the major contributors to biogeochemical cycling in these soils. For example, while similar to other soils and to each other, the microbial communities from 29 heath tundra sites across the Arctic were also heterogeneous and distinctive, as 55% of sequences at each site were unique to that site (Chu et al., 2010). Again, while similar overall, important genera of both Bacteria and Archaea were unique to each site in a study of four low-center polygons in the western Canadian Arctic (Frank-Fahle et al., 2014). These site-specific communities likely form in response to variations in local conditions often due to variations in micro-topography (see earlier). Indeed, a study of Arctic and Antarctic soils found that community structure was affected by connectivity of landforms (Ferrari et al., 2016). If soils were geologically connected, community composition was shaped by distance and landform (aspect, slope, elevation) which together create gradients in local soil chemical environments. If soils were fragmented, local soil conditions (particularly Cl) most influenced community composition.

Furthermore, numerous studies have demonstrated that microbial communities vary by depth in Arctic soils. For example, in moist acidic tundra with dwarf shrubs of the Seward Peninsula, Alaska, the bacterial diversity, total P, and C:N ratio decreased with depth, while pH and SOM decomposition increased with depth (Kim et al., 2016). Interestingly, community composition correlated with pH and total P, while the composition of functional genes only correlated with pH. The bacterial community shifted from being dominated by Alphaproteobacteria and Acidobacteria at the surface to being dominated by Actinobacteria, Betaproteobacteria, and Chloroflexi at depth. Similar results have been documented in an acidic wetland and mineral cryosols of the High Arctic (Wilhelm et al., 2011; Yergeau et al., 2010), low center polygons in the western Canadian Arctic and Alaska (Frank-Fahle et al., 2014; Lipson et al., 2015), as well as in graminoid and shrub-dominated tundra of Alaska (Deng et al., 2015). Similarly, polar desert soils from Svalbard were dominated by Proteobacteria, Actinobacteria, Chloroflexi, Acidobacteria, and Bacteroidetes where composition and diversity correlated with pH, Ca, and P (McCann et al., 2016). While there are few studies of soils from Greenland, a recent study demonstrated that microbial communities of these polar desert soils varied by depth, correlated with pH, and were dominated by Acidobacteria, Actinobacteria, and Bacteroidetes (Ganzert et al., 2014). Even in cryoturbated shrubby tundra along the Kolyma River, Siberia, diversity decreased with depth, except for where buried topsoils were found, and community composition correlated with pH, C:N ratio, TOC, and moisture (Gittel et al., 2014).

Predictably, microbial communities of Arctic soils are highly resistant to freeze-thaw cycles. For example, microbial communities from Finnish tundra soil changed very little during and after exposure to repeated freeze-thaw (Mannisto et al., 2009). Resilience to freeze-thaw was also evident in microbial communities from active layer tundra soils of Svalbard in which the abundance of many OTUs did not change seasonally (Schostag et al., 2015). In addition, bacteria appeared to remain active (based on very little seasonal change in the ratio of transcripts to gene copy number) when frozen during the winter. Likewise, repeated freeze-thaw had no significant effect on the mineralization of low molecular weight DOC in two tundra soils from Svalbard (Foster et al., 2016). In contrast, microbial communities from temperate soils change significantly after repeated freeze-thaw (Foster et al., 2016; Stres et al., 2010). Interestingly, microbial communities from soils of the Low Arctic and the High Arctic changed significantly in response to freeze-thaw (and was evident as changes in abundance and decreased diversity), while communities from Subarctic soils were found to change very little (Kumar et al., 2013). Hence, the resilience of microbial communities to repeated freeze-thaw appears dependent on their in situ exposure to freeze-thaw conditions, which are more frequent in the Subarctic.

3.4 Summary

Arctic soils provide a unique landscape for the examination of biogeochemistry as affected by low temperatures. Biogeochemical trends include a north-south gradient, a mosaic of site-specific conditions in part created by the presence of permafrost, and distinct vertical layers. Low temperatures further affect the rates of biogeochemical processes by slowing organic matter decomposition and through the mechanical action of freeze-thaw cycles that alter soil structure and affect burial of organic matter. Arctic soils contain microbial communities that are generally similar to microbial communities from other soils. Given that microorganisms are often the major contributors to biogeochemical cycling in Arctic soils, it is not surprising to find that the structure of microbial communities (i.e., a mosaic of site-specific communities and vertical layers) in Arctic soils is similar to trends seen in the biogeochemistry of these soils.

4 Additional constraints of permafrost
4.1 Life in a frozen matrix

For microorganisms, the defining character of permafrost is its frozen state, which presents organisms with the combined stresses of low temperatures and of living in a confined matrix. In this frozen matrix, liquid water is limited and confined to thin films between ice crystals and minerals. The amount of liquid water is influenced by the solute concentrations and can be higher in soils with more solutes. For example, when marine sediments freeze, their high solute concentration can prompt the formation of brine lenses known as cryopegs within the resultant permafrost (Colangelo-Lillis et al., 2016; Gilichinsky et al., 2003). Small soil particles can also increase the liquid water content through surface interactions and contributes to the nearly 10% unfrozen water within Arctic loamy soils at temperatures of $-9°C$ to $-12°C$ (Steven et al., 2006). At temperatures of $-15°C$ and below, liquid water is probably limited to a few layers of adsorbed water (Goordial et al., 2016a; Jakosky et al., 2003) which are not likely to be accessible to microorganisms (Stevenson et al., 2015).

Confinement to a small space such as a thin film between ice crystals and soil particles leads not just to the reduced availability of water, but also to the reduced diffusion of nutrients. Without diffusion of nutrients, microorganisms are not likely to be metabolically active. Indeed, when a limit on substrate availability due to low diffusion rates through thin films in frozen soils was included during modeling of carbon respiration in soils, the long-term storage of carbon in permafrost was more accurately predicted (Schaefer and Jafarov, 2016). This lack of microbial activity is further supported by the high lability of carbon upon thawing which suggests that permafrost carbon has not been noticeably decomposed by microorganisms while frozen (Schuur et al., 2008; Waldrop et al., 2010). For example, when the organic-rich Yedoma permafrost from northeast Siberia was examined over an age range of 30.1 to >55 thousand years old there was no correlation between any indicator of organic matter quality (higher plant fatty acids, C/N ratio, $\delta^{13}C$ values of TOC, or carbon preference index) and depth or age indicating that no significant microbial decomposition had occurred over this time frame (Strauss et al., 2015). Other studies of frozen soils confirm that microbial respiration effectively ceases at temperatures of $-7°C$ to $-8°C$ (Schaefer and Jafarov, 2016).

A lack of nutrients leads to starvation stress that cells can contend with by entering a dormant or a semidormant state to remain viable in the long term. A semidormant state would include intermittent metabolism for maintenance and repair of accumulated damage. Microorganisms have been isolated from permafrost environments in many studies (Gilichinsky and Wagener, 1995; Johnson et al., 2007; Vishnivetskaya et al., 2006), demonstrating that survival of these long-term low temperature conditions is frequently realized and likely required intermittent maintenance and repair for microorganisms to remain viable. Indeed, long-term burial in permafrost appears to select against endospores which have no maintenance or repair metabolism (Johnson et al., 2007). And transcriptomic analysis of permafrost from moist acidic tundra near Toolik, Alaska demonstrated that genes expressed in permafrost primarily reflected maintenance and dormancy in response to stress (Coolen and Orsi, 2015). In addition, various metabolic activities have been demonstrated in soil microcosms at low temperatures (Nikrad et al., 2016), also suggesting the ability for long-term survival that requires metabolism in permafrost.

4.2 Additional stresses

As long as microbes can maintain some intermittent metabolism for repair of accumulated damage, other stresses that might be present in permafrost do not seem to have a large impact on individual microorganisms or communities. For example, permafrost contains radioactive minerals that can cause damage to microorganism at an estimated rate of 2 mGy per year which would result in a lethal dose after 10 million years (McKay, 2001). While DNA damage appears to increase with age of permafrost (Hansen et al., 2006), repair of DNA damage is also evident (Johnson et al., 2007). Rates of DNA synthesis by the permafrost isolate *Psychrobacter arcticus* 273-4 while frozen at −15°C are high enough to offset the expected DNA damage from radiation in permafrost (Amato et al., 2010). In addition, *P. arcticus* 273-4 can repair double strand breaks (DSB) of DNA while frozen at −15°C (Dieser et al., 2013). The average rate of repair was estimated at 8 DSBs per year which is far greater than the estimated rate of damage due to radiation (1 DSB every 14,500 years), and the authors concluded that DSBs should not be an issue for long-term survival if energy is available. (This suggests that DNA damage seen in DNA extracted from permafrost is from completely dormant or dead organisms.) Over time, chemical degradation like nucleotide depurination and amino acid racemization can also accumulate, but again, in permafrost rates of damage are probably lower than rates of repair. In deep subseafloor, sediments' metabolic rates are also very low, but still $10\times$ faster than required to offset damage due to racemization and $1000\times$ faster than required to offset damage due to depurination (Hoehler and Jorgensen, 2013). Cold environments like permafrost will have lower rates of damage due to lower temperatures. Indeed, deracemization of amino acids (via microbial enzymes) has been documented in Siberian permafrost up to 25,000 years old (Brinton et al., 2002) demonstrating that limited intermittent metabolism is sufficient to counteract accumulated damage in permafrost and facilitate long-term survival.

Subfreezing temperatures create additional abiotic stresses on cells in permafrost, such as osmotic stress due to high solute concentrations in the remaining liquid between ice crystals (due to solute exclusion from the ice matrix during freezing) and soil particles. As a result, many organisms isolated from permafrost are halotolerant (e.g., Bakermans et al., 2006; Rodrigues et al., 2006). Halophiles are not typically isolated from permafrost, but may be isolated from cryopegs (Gilichinsky et al., 2005; Shcherbakova et al., 2013) and hypersaline springs (Lamarche-Gagnon et al., 2015) found in permafrost.

4.3 Microbial communities

Permafrost usually contains a lower number of cells than the active layer. For example, fewer rRNA gene copies were present in permafrost underlying shrubby tundra than in topsoil of the active layer along the Kolyma River, Russia (Gittel et al., 2014); fewer transcripts were recovered from permafrost than from the active layer at Bonanza Creek, Fairbanks, AK (Hultman et al., 2015); and fewer genes and taxa were detected via qPCR in permafrost than the active layer near Eureka, Ellesmere Island, CAN (Yergeau et al., 2010). In addition, the microbial communities of permafrost are distinct from the active layer with representative phyla including the Proteobacteria, Firmicutes, Chloroflexi, Acidobacteria, Actinobacteria, and Bacteroidetes (Jansson and Tas, 2014; Rivkina et al., 2016; Yergeau et al., 2010). Not surprisingly, dominant phyla vary due to the heterogeneity of permafrost.

For example, Actinobacteria, Alphaproteobacteria, Acidobacteria, and Gammaproteobacteria dominated in permafrost underlying shrubby tundra along the Kolyma River (Gittel et al., 2014). In permafrost from Bonanza Creek, Chloroflexi, Proteobacteria, and Actinobacteria dominated as determined by numbers of 16S rRNA genes; however, high ratios of transcripts to genes in Proteobacteria, Acidobacteria, and Firmicutes suggested these phyla were highly active in the permafrost (Hultman et al., 2015). A different study of Alaskan permafrost also found high levels of transcripts from Proteobacteria, Firmicutes, Acidobacteria, and Actinobacteria suggesting high levels of activity of these phyla in permafrost underlying tussock-sedge tundra near Toolik Lake, AK (Coolen and Orsi, 2015). As in other soils, community composition of permafrost is affected by abiotic factors such as pH and C:N ratio (Gittel et al., 2014); however, the paucity of studies examining the effect of abiotic factors on microbial community composition limits interpretation. Indeed, a study of permafrost and an adjacent thermokarst bog in interior Alaska concluded that "global molecular data were a poorer predictor of biogeochemical process rates in permafrost than the other soils" (Hultman et al., 2015).

4.4 Summary

The low temperatures and frozen state of permafrost provide an excellent model for long-term maintenance and survival of microorganisms that results in long-term storage of carbon. In combination with the very low temperatures (that lead to low rates of metabolism), other stresses do not appear to have a major impact on organisms or communities (few halophiles and psychrophiles). Microbial communities in permafrost are very much like the communities in the active layer in terms of abilities (such as growth rates, optimum growth temperatures, substrate utilization) suggesting very little adaptation to permanently frozen conditions and the continued effect of selection for adaptation to freeze-thaw conditions in these communities (Ernakovich and Wallenstein, 2015).

5 Anticipated changes with warming

Permafrost in the Arctic currently represents a carbon sink of $\sim$1670 Pg C, some 10–30 times the amount found in other deep soils (Mueller et al., 2015; Tarnocai et al., 2009). This organic material persists primarily due to the constant subzero temperatures, low liquid water content, and anoxic conditions that all contribute to slowing the metabolic activity of microorganisms to extremely low levels (see earlier). The contribution of permafrost to global carbon cycling is poised to change as the Arctic warms due to global climate change and the cold-adapted microorganisms inhabiting these environments become more active.

In the Arctic, the average annual surface temperature has increased almost 1.5°C since 1950 (Jeffries et al., 2014) and is projected to rise a total of 7–8°C by 2100 (IPCC, 2007). Concomitant with increases in air temperature, the temperatures of permafrost have increased by 0.5–2°C since the 1970s, the thickness of the active layer has increased by up to 90 cm, and local degradation (thawing) of permafrost is evident throughout the Arctic from Alaska to Siberia (IPCC, 2013; Payette et al., 2004; Romanovsky et al., 2010; Streletskiy et al., 2015). The increase in temperature has dramatic effects on the metabolic activity of microorganisms present: every 10°C rise in temperature increases enzymatic and metabolic rates (Q_{10} values) by two to three times or more (Mikan et al., 2002; Oquist et al., 2009). Indeed, studies have shown a rapid response of microorganisms to warming and thawing. Both the community and functional gene composition changed rapidly upon thaw of permafrost from lowland soil in Hess Creek, Alaska (Mackelprang et al., 2011). Furthermore, while gene expression in permafrost from moist acidic tundra near Toolik, Alaska primarily reflected maintenance and dormancy in response to stress, gene expression in thawed permafrost was dominated by enzymes for the decomposition of SOM (Coolen and Orsi, 2015). Changes in microbial communities are also evident at a longer time scale. For example, along a thaw gradient in Alaska (from minimal thaw to thaw since before 1955), differences in OTU abundance were correlated with extent of thaw (Deng et al., 2015). Notably, Actinobacteria, known for degradation of complex recalcitrant carbon sources, increased with extent of thaw at the lowest depths. In addition, respiration increased by 39% and respiration of old carbon doubled in extensively thawed permafrost sites (Deng et al., 2015). Gas flux experiments have also demonstrated that thawing permafrost releases "old" organic material at significant rates on time scales of 15 to 100 years (Coolen et al., 2011; Schuur et al., 2009; Shaver et al., 2006).

Microbial activity (and hence SOM mineralization) is not only increased directly, but also indirectly, by warmer temperatures. For example, increased soil temperature and active layer thickness may increase weathering, resulting in higher nutrient availability and greater concentrations of exchangeable base cations (Schuur et al., 2008). In addition, a warmer climate may also result in a more nutrient-rich active layer as increased cryoturbation and permafrost thaw distribute macronutrients and previously frozen, unavailable organic matter upward (Miller et al., 2015). Alternatively, thaw may increase soil saturation and flooding, increasing organic matter accumulation and nutrient sequestering (Whittinghill and Hobbie, 2012). The heterogeneity of soils in the Arctic and the limited number of soil types examined to date leads to variation in results and makes extrapolation across the entire Arctic difficult. A recent study attempted to identify general responses of Arctic soils and demonstrated that a 10°C increase in temperature resulted in 2× more C released regardless of soil type (Schadel et al., 2016). In addition, oxic conditions resulted in 3.4× more C released than anoxic conditions regardless of incubation temperature and soil type.

However, large uncertainties remain in the total amount of C release to be expected as permafrost thaws due to the heterogeneity of the mosaic landscape of the Arctic. In addition, the effect of changes in thaw, moisture content, vegetation, and other disturbances on microbial activity are still not well understood and limit the ability to accurately predict and model nutrient cycles in the Arctic (Mackelprang et al., 2016). Continued study will help to define the role of Arctic soils in carbon cycling and global climate regulation in the future.

6 Conclusions

The terrestrial Arctic provides a model system for the study of growth and survival of microorganisms at low temperatures. The effect of low temperature and other abiotic factors on microbial activity is evident in the biogeochemical trends that emerge. In particular, the mosaic of site-specific conditions created in part by the presence of permafrost is unique and contributes to the site-specific biogeochemical regimes and microbial communities that develop. As exposure to freeze-thaw increases, microbial communities from Arctic soils are increasingly resilient to freeze-thaw conditions. In addition, most isolates from Arctic soils are eurypsychrophiles that may have some unique cold adaptations (e.g., isozymes) and can serve as model organisms for living across a broad temperature range that includes subfreezing temperatures. Low temperatures further affect the rates of biogeochemical processes by slowing organic matter decomposition.

The frozen state of permafrost provides an excellent model for long-term maintenance and survival of microorganisms that results in the long-term storage of carbon. Very low levels of maintenance metabolism facilitate the survival of bacteria trapped in permafrost for thousands to millions of years. The abilities of microorganisms in permafrost are comparable to those from the active layer suggesting little adaptation to permanently frozen conditions and the continued effect of selection for adaptation to freeze-thaw conditions in these communities. As the temperature of permafrost and the active layer increases, the microorganisms present respond rapidly leading to the conversion of previously stored organic carbon to metabolic waste products like CO_2 and CH_4. The extent of organic matter decomposition will depend on many factors; more study is needed to better understand how thawing permafrost will affect carbon cycling in the Arctic and global climate regulation.

Lastly, the long-term low temperature conditions of the Arctic provide model systems for understanding the survival of life elsewhere in the solar system. Permafrost is a good model for long-term low temperature survival, while the mineral soils of the High Arctic may be a model for near surface ground and permafrost on Mars. In addition, hypersaline springs from deep aquifers within the High Arctic may be relevant models for recurring slope lineae on Mars. Continued study of these habitats will expand our knowledge on the limits and capabilities of terrestrial life under extreme conditions for extrapolation to potentially habitable conditions found beyond Earth.

References

Amato, P., Doyle, S.M., Battista, J.R., Christner, B.C., 2010. Implications of subzero metabolic activity on long-term microbial survival in terrestrial and extraterrestrial permafrost. Astrobiology 10, 789–798.

Anisimov, O.A., Shiklomanov, N.I., Nelson, F.E., 1997. Global warming and active-layer thickness: results from transient general circulation models. Glob. Planet. Chang. 15, 61–77.

Bakermans, C., 2012. Psychrophiles: life in the cold. In: Anitori, R. (Ed.), Extremophiles: Microbiology and Biotechnology. Horizon Scientific Press, Hethersett, England, UK, pp. 53–76.

Bakermans, C., Ayala-del-Río, H.L., Ponder, M.A., Vishnivetskaya, T., Gilichinsky, D., Thomashow, M.F., Tiedje, J.M., 2006. *Psychrobacter cryohalolentis* sp. nov. and *Psychrobacter arcticus* sp. nov. isolated from Siberian permafrost. Int. J. Syst. Evol. Microbiol. 56, 1285–1291.

Bergholz, P.W., Bakermans, C., Tiedje, J.M., 2009. *Psychrobacter arcticus* 273-4 uses resource efficiency and molecular motion adaptations for subzero temperature growth. J. Bacteriol. 191, 2340–2352.

Blaud, A., Lerch, T.Z., Phoenix, G.K., Osborn, A.M., 2015. Arctic soil microbial diversity in a changing world. Res. Microbiol. 166, 796–813.

Bockheim, J.G., 2015. Global distribution of cryosols with mountain permafrost: an overview. Permafr. Periglac. Process. 26, 1–12.

Bockheim, J.G., Walker, D.A., Everett, L.R., Nelson, F.E., Shiklomanov, N.I., 1998. Soils and cryoturbation in moist nonacidic and acidic tundra in the Kuparuk River basin, Arctic Alaska. USA. Arct. Alp. Res. 30, -174.

Brinton, K., Tsapin, A., Gilichinsky, D., McDonald, G., 2002. Aspartic acid racemization and age-depth relationships for organic carbon in Siberian permafrost. Astrobiology 2, 77–82.

Brown, J., Romanovsky, V.E., 2008. Report from the International Permafrost Association: state of permafrost in the first decade of the 21st century. Permafr. Periglac. Process. 19, 255–260.

Buckeridge, K.M., Grogan, P., 2010. Deepened snow increases late thaw biogeochemical pulses in mesic low arctic tundra. Biogeochemistry 101, 105–121.

Christiansen, C.T., Haugwitz, M.S., Prieme, A., Nielsen, C.S., Elberling, B., Michelsen, A., Grogan, P., Blok, D., 2017. Enhanced summer warming reduces fungal decomposer diversity and litter mass loss more strongly in dry than in wet tundra. Glob. Chang. Biol. 23, 406–420.

Chu, H., Fierer, N., Lauber, C.L., Caporaso, J.G., Knight, R., Grogan, P., 2010. Soil bacterial diversity in the Arctic is not fundamentally different from that found in other biomes. Environ. Microbiol. 12, 2998–3006.

Colangelo-Lillis, J., Eicken, H., Carpenter, S.D., Deming, J.W., 2016. Evidence for marine origin and microbial-viral habitability of sub-zero hypersaline aqueous inclusions within permafrost near Barrow, Alaska. FEMS Microbiol. Ecol. 92. fiw053.

Coolen, M.J.L., Orsi, W.D., 2015. The transcriptional response of microbial communities in thawing Alaskan permafrost soils. Front. Microbiol. 6.

Coolen, M.J.L., van de Giessen, J., Zhu, E.Y., Wuchter, C., 2011. Bioavailability of soil organic matter and microbial community dynamics upon permafrost thaw. Environ. Microbiol. https://doi.org/10.1111/j.1462-2920.2011.02489.x.

Davidson, E.A., Janssens, I.A., 2006. Temperature sensitivity of soil carbon decomposition and feedbacks to climate change. Nature 440, 165–173.

De Maayer, P., Anderson, D., Cary, C., Cowan, D.A., 2014. Some like it cold: understanding the survival strategies of psychrophiles. EMBO Rep. 15, 508–517.

Deng, J., Gu, Y., Zhang, J., Xue, K., Qin, Y., Yuan, M., Yin, H., He, Z., Wu, L., Schuur, E.A.G., et al., 2015. Shifts of tundra bacterial and archaeal communities along a permafrost thaw gradient in Alaska. Mol. Ecol. 24, 222–234.

Dieser, M., Battista, J.R., Christner, B.C., 2013. DNA double-strand break repair at −15 °C. Appl. Environ. Microbiol. 79, 7662–7668.

Edwards, K.A., Jefferies, R.L., 2013. Inter-annual and seasonal dynamics of soil microbial biomass and nutrients in wet and dry low-Arctic sedge meadows. Soil Biol. Biochem. 57, 83–90.

Ernakovich, J.G., Wallenstein, M.D., 2015. Permafrost microbial community traits and functional diversity indicate low activity at in situ thaw temperatures. Soil Biol. Biochem. 87, 78–89.

Feller, G., 2013. Psychrophilic enzymes: from folding to function and biotechnology. Scientifica (Cairo) 2013, 512840.

Ferrari, B.C., Bissett, A., Snape, I., van Dorst, J., Palmer, A.S., Ji, M., Siciliano, S.D., Stark, J.S., Winsley, T., Brown, M.V., 2016. Geological connectivity drives microbial community structure and connectivity in polar, terrestrial ecosystems. Environ. Microbiol. 18, 1834–1849.

Fierer, N., Jackson, R.B., 2006. The diversity and biogeography of soil bacterial communities. Proc. Natl. Acad. Sci. U. S. A. 103, 626–631.

Finster, K.W., Herbert, R.A., Lomstein, B.A., 2009. *Spirosoma spitsbergense* sp nov and *Spirosoma luteum* sp nov., isolated from a high Arctic permafrost soil, and emended description of the genus *Spirosoma*. Int. J. Syst. Evol. Microbiol. 59, 839–844.

Foster, A., Jones, D.L., Cooper, E.J., Roberts, P., 2016. Freeze–thaw cycles have minimal effect on the mineralisation of low molecular weight, dissolved organic carbon in Arctic soils. Polar Biol. 39, 2387–2401.

Frank-Fahle, B.A., Yergeau, É., Greer, C.W., Lantuit, H., Wagner, D., 2014. Microbial functional potential and community composition in permafrost-affected soils of the NW Canadian Arctic. PLoS ONE 9, e84761.

Froese, D.G., Westgate, J.A., Reyes, A.V., Enkin, R.J., Preece, S.J., 2008. Ancient permafrost and a future, warmer Arctic. Science 321, 1648.

Gallardo, K., Candia, J.E., Remonsellez, F., Escudero, L.V., Demergasso, C.S., 2016. The ecological coherence of temperature and salinity tolerance interaction and pigmentation in a non-marine vibrio isolated from Salar de Atacama. Front. Microbiol. 7.

Ganzert, L., Bajerski, F., Wagner, D., 2014. Bacterial community composition and diversity of five different permafrost-affected soils of Northeast Greenland. FEMS Microbiol. Ecol. 89, 426–441.

Gerday, C., 2011. Life at the Extremes of Temperature. Amer Soc Microbiology, Washington.

Gerday, C., 2013. Psychrophily and catalysis. Biology (Basel) 2, 719–741.

Gilichinsky, D., Wagener, S., 1995. Microbial life in permafrost: a historical review. Permafrost Periglac 6, 243–250.

Gilichinsky, D., Rivkina, E., Shcherbakova, V., Laurinavichuis, K., Tiedje, J., 2003. Supercooled water brines within permafrost—an unknown ecological niche for microorganisms: a model for astrobiology. Astrobiology 3, 331–341.

Gilichinsky, D., Rivkina, E., Bakermans, C., Shcherbakova, V., Petrovskaya, L., Ozerskaya, S., Ivanushkina, N., Kochkina, G., Laurinavichuis, K., Pecheritsina, S., et al., 2005. Biodiversity of cryopegs in permafrost. FEMS Microbiol. Ecol. 53, 117–128.

Gittel, A., Barta, J., Kohoutova, I., Mikutta, R., Owens, S., Gilbert, J., Schnecker, J., Wild, B., Hannisdal, B., Maerz, J., et al., 2014. Distinct microbial communities associated with buried soils in the Siberian tundra. ISME J. 8, 841–853.

Goordial, J., Davila, A., Lacelle, D., Pollard, W., Marinova, M.M., Greer, C.W., DiRuggiero, J., McKay, C.P., Whyte, L.G., 2016a. Nearing the cold-arid limits of microbial life in permafrost of an upper dry valley, Antarctica. ISME J. 10, 1613–1624.

Goordial, J., Raymond-Bouchard, I., Riley, R., Ronholm, J., Shapiro, N., Woyke, T., LaButti, K.M., Tice, H., Amirebrahimi, M., Grigoriev, I.V., et al., 2016b. Improved high-quality draft genome sequence of the eurypsychrophile *Rhodotorula* sp. JG1b, isolated from permafrost in the hyperarid upper-elevation McMurdo Dry Valleys, Antarctica. Genome Announc. 4.

Gray, N.D., McCann, C.M., Christgen, B., Ahammad, S.Z., Roberts, J.A., Graham, D.W., 2014. Soil geochemistry confines microbial abundances across an arctic landscape; implications for net carbon exchange with the atmosphere. Biogeochemistry 120, 307–317.

Hansen, A.J., Mitchell, D.L., Wiuf, C., Panikert, L., Brand, T.B., Binladen, J., Gilichinsky, D.A., Ronn, R., Willerslev, E., 2006. Crosslinks rather than strand breaks determine access to ancient DNA sequences from frozen sediments. Genetics 173, 1175–1179.

Herndon, E.M., Yang, Z.M., Bargar, J., Janot, N., Regier, T., Graham, D., Wullschleger, S., Gu, B.H., Liang, L.Y., 2015. Geochemical drivers of organic matter decomposition in arctic tundra soils. Biogeochemistry 126, 397–414.

Hobbie, S.E., Gough, L., 2002. Foliar and soil nutrients in tundra on glacial landscapes of contrasting ages in northern Alaska. Oecologia 131, 453–462.

Hoehler, T.M., Jorgensen, B.B., 2013. Microbial life under extreme energy limitation. Nat. Rev. Microbiol. 11, 83–94.

Hultman, J., Waldrop, M.P., Mackelprang, R., David, M.M., McFarland, J., Blazewicz, S.J., Harden, J., Turetsky, M.R., McGuire, A.D., Shah, M.B., et al., 2015. Multi-omics of permafrost, active layer and thermokarst bog soil microbiomes. Nature 521, 208–212.

Husson, O., 2013. Redox potential (Eh) and pH as drivers of soil/plant/microorganism systems: a transdisciplinary overview pointing to integrative opportunities for agronomy. Plant Soil 362, 389–417.

IPCC, 2007. Climate Change 2007: The Physical Science Basis. Contribution of Working Group I to the Fourth Assessment Report of the Intergovernmental Panel on Climate Change. Cambridge University Press, New York.

IPCC, 2013. Stocker, T.F., Qin, D., Plattner, G.-K., Tignor, M., Allen, S.K., Boschung, J., Nauels, A., Xia, Y., Bex, V., Midgley, P.M. (Eds.), Climate change 2013: the physical science basis. Contribution of Working Group I to the Fifth Assessment Report of the Intergovernmental Panel on Climate Change. Cambridge University Press, Cambridge, United Kingdom/New York, NY.

Iturrate-Garcia, M., O'Brien, M.J., Khitun, O., Abiven, S., Niklaus, P.A., Schaepman-Strub, G., 2016. Interactive effects between plant functional types and soil factors on tundra species diversity and community composition. Ecol. Evol. 6, 8126–8137.

Jakosky, B.M., Nealson, K.H., Bakermans, C., Ley, R.E., Mellon, M.T., 2003. Subfreezing activity of microorganisms and the potential habitability of Mars' polar regions. Astrobiology 3, 343–350.

Jansson, J.K., Tas, N., 2014. The microbial ecology of permafrost. Nat. Rev. Microbiol. 12, 414–425.

Jeffries, M.O., Richter-Menge, J., Overland, J.E., 2014. Arctic Report Card 2014. http://www.arctic.noaa.gov/reportcard.

Johnson, S., Hebsgaard, M., Christensen, T., Mastepanov, M., Nielsen, R., Munch, K., Brand, T., Gilbert, M., Zuber, M., Bunce, M., et al., 2007. Ancient bacteria show evidence of DNA repair. PNAS 104, 14401–14405.

Jonasson, S., Michelsen, A., Schmidt, I.K., Nielsen, E.V., Callaghan, T.V., 1996. Microbial biomass C, N and P in two arctic soils and responses to addition of NPK fertilizer and sugar: implications for plant nutrient uptake. Oecologia 106, 507–515.

Jorgensen, C.J., Johansen, K.M.L., Westergaard-Nielsen, A., Elberling, B., 2015. Net regional methane sink in High Arctic soils of northeast Greenland. Nat. Geosci. 8, 20–23.

Keller, K., Blum, J.D., Kling, G.W., 2007. Geochemistry of soils and streams on surfaces of varying ages in arctic Alaska. Arct. Antarct. Alp. Res. 39, 84–98.

Kelley, A.M., Epstein, H.E., Ping, C.L., Walker, D.A., 2012. Soil nitrogen transformations associated with small patterned-ground features along a North American Arctic Transect. Permafr. Periglac. Process. 23, 196–206.

Kim, H.M., Lee, M.J., Jimg, J.Y., Hwang, C.Y., Kim, M., Ro, H.M., Chun, J., Lee, Y.K., 2016. Vertical distribution of bacterial community is associated with the degree of soil organic matter decomposition in the active layer of moist acidic tundra. J. Microbiol. 54, 713–723.

Kokelj, S.V., Burn, C.R., 2005. Geochemistry of the active layer and near-surface permafrost, Mackenzie delta region, Northwest Territories, Canada. Can. J. Earth Sci. 42, 37–48.

Kumar, N., Grogan, P., Chu, H., Christiansen, C., Walker, V., 2013. The effect of freeze-thaw conditions on Arctic soil bacterial communities. Biology 2, 356.

Lamarche-Gagnon, G., Comery, R., Greer, C.W., Whyte, L.G., 2015. Evidence of in situ microbial activity and sulphidogenesis in perennially sub-0°C and hypersaline sediments of a high Arctic permafrost spring. Extremophiles 19, 1–15.

Lipson, D.A., Jha, M., Raab, T.K., Oechel, W.C., 2010. Reduction of iron (III) and humic substances plays a major role in anaerobic respiration in an Arctic peat soil. J. Geophys. Res. Biogeosci. 115.

Lipson, D.A., Zona, D., Raab, T.K., Bozzolo, F., Mauritz, M., Oechel, W.C., 2012. Water-table height and microtopography control biogeochemical cycling in an Arctic coastal tundra ecosystem. Biogeosciences 9, 577–591.

Lipson, D.A., Raab, T.K., Goria, D., Zlamal, J., 2013. The contribution of Fe(III) and humic acid reduction to ecosystem respiration in drained thaw lake basins of the Arctic Coastal Plain. Glob. Biogeochem. Cycles 27, 399–409.

Lipson, D.A., Raab, T.K., Parker, M., Kelley, S.T., Brislawn, C.J., Jansson, J., 2015. Changes in microbial communities along redox gradients in polygonized Arctic wet tundra soils. Environ. Microbiol. Rep. 7, 649–657.

Mackelprang, R., Waldrop, M.P., DeAngelis, K.M., David, M.M., Chavarria, K.L., Blazewicz, S.J., Rubin, E.M., Jansson, J.K., 2011. Metagenomic analysis of a permafrost microbial community reveals a rapid response to thaw. Nature 480, 368–371.

Mackelprang, R., Saleska, S.R., Jacobsen, C.S., Jansson, J.K., Tas, N., 2016. Permafrost meta-omics and climate change. In: Jeanloz, R., Freeman, K.H. (Eds.), Annual Review of Earth and Planetary Sciences.In: vol. 44. pp. 439–462.

Mannisto, M.K., Tiirola, M., Haggblom, M.M., 2009. Effect of freeze-thaw cycles on bacterial communities of Arctic tundra soil. Microb. Ecol. 58, 621–631.

Margesin, R., 2017. Psyhcrophiles: from biodiversity to biotechnology, second ed. Springer International Publishing.

McCann, C.M., Wade, M.J., Gray, N.D., Roberts, J.A., Hubert, C.R.J., Graham, D.W., 2016. Microbial communities in a high Arctic polar desert landscape. Front. Microbiol. 7.

McKay, C., 2001. The deep biosphere: lessons for planetary exploration. In: Fredrickson, J., Fletcher, M. (Eds.), Subsurface Microbiology and Biogeochemistry. Wiley-Liss, New York, pp. 315–328.

Mikan, C.J., Schimel, J.P., Doyle, A.P., 2002. Temperature controls of microbial respiration in arctic tundra soils above and below freezing. Soil Biol. Biochem. 34, 1785–1795.

Miller, K.E., Lai, C.T., Friedman, E.S., Angenent, L.T., Lipson, D.A., 2015. Methane suppression by iron and humic acids in soils of the Arctic Coastal Plain. Soil Biol. Biochem. 83, 176–183.

Mueller, C.W., Rethemeyer, J., Kao-Kniffin, J., Loppmann, S., Hinkel, K.M., Bockheim, J., 2015. Large amounts of labile organic carbon in permafrost soils of northern Alaska. Glob. Chang. Biol.

Mykytczuk, N.C.S., Wilhelm, R., Whyte, L.G., 2012. *Planococcus halocryophilus* sp. nov.; an extreme subzero species from high Arctic permafrost. Int. J. Syst. Evol. Microbiol. 62, 1937–1944.

Nadelhoffer, K.J., Johnson, L., Laundre, J., Giblin, A.E., Shaver, G.R., 2002. Fine root production and nutrient content in wet and moist arctic tundras as influenced by chronic fertilization. Plant Soil 242, 107–113.

Niederberger, T.D., Steven, B., Charvet, S., Barbier, B., Whyte, L.G., 2009. *Virgibacillus arcticus* sp. nov., a moderately halophilic, endospore-forming bacterium from permafrost in the Canadian high Arctic. Int. J. Syst. Evol. Microbiol. 59, 2219–2225.

Nikrad, M.P., Kerkhof, L.J., Haggblom, M.M., 2016. The subzero microbiome: microbial activity in frozen and thawing soils. FEMS Microbiol. Ecol. 92.

Oquist, M.G., Sparrman, T., Klemedtsson, L., Drotz, S.H., Grip, H., Schleucher, J., Nilsson, M., 2009. Water availability controls microbial temperature responses in frozen soil CO_2 production. Glob. Chang. Biol. 15, 2715–2722.

Payette, S., Delwaide, A., Caccianiga, M., Beauchemin, M., 2004. Accelerated thawing of sub-arctic peatland permafrost over the last 50 years. Geophys. Res. Lett. 31, 4.

Pearce, A.R., Rastetter, E.B., Kwiatkowski, B.L., Bowden, W.B., Mack, M.C., Jiang, Y., 2015. Recovery of arctic tundra from thermal erosion disturbance is constrained by nutrient accumulation: a modeling analysis. Ecol. Appl. 25, 1271–1289.

Penton, C.R., Yang, C.Y., Wu, L.Y., Wang, Q., Zhang, J., Liu, F.F., Qin, Y.J., Deng, Y., Hemme, C.L., Zheng, T.L., et al., 2016. NifH-harboring bacterial community composition across an Alaskan permafrost thaw gradient. Front. Microbiol. 7.

Prescott, C.E., Weetman, G.F., Demontigny, L.E., Preston, C.M., Keenan, R.J., 1995. Carbon chemistry and nutrient supply in cedar-hemlock and hemlock-amabilis fir forest floors. In: McFee, W.W., Kelly, J.M., Bigham, J.M. (Eds.), Carbon Forms and Functions in Forest Soils. Soil Science Society of America, Inc. USA, Madison, WI, pp. 377–396.

Przybylak, R., 2000. Temporal and spatial variation of surface air temperature over the period of instrumental observations in the Arctic. Int. J. Climatol. 20, 587–614.

Reyes, F.R., Lougheed, V.L., 2015. Rapid nutrient release from permafrost thaw in arctic aquatic ecosystems. Arct. Antarct. Alp. Res. 47, 35–48.

Rivkina, E., Petrovskaya, L., Vishnivetskaya, T., Krivushin, K., Shmakova, L., Tutukina, M., Meyers, A., Kondrashov, F., 2016. Metagenomic analyses of the late Pleistocene permafrost—additional tools for reconstruction of environmental conditions. Biogeosciences 13, 2207–2219.

Rodrigues, D.F., Goris, J., Vishnivetskaya, T., Gilichinsky, D., Thomashow, M.F., Tiedje, J.M., 2006. Characterization of *Exiguobacterium* isolates from the Siberian permafrost. Description of *Exiguobacterium sibiricum* sp nov. Extremophiles 10, 285–294.

Romanovsky, V.E., Drozdov, D.S., Oberman, N.G., Malkova, G.V., Kholodov, A.L., Marchenko, S.S., Moskalenko, N.G., Sergeev, D.O., Ukraintseva, N.G.,

Abramov, A.A., et al., 2010. Thermal state of permafrost in Russia. Permafr. Periglac. Process. 21, 136–155.

Rousk, K., Sorensen, P.L., Michelsen, A., 2016. Nitrogen transfer from four nitrogen-fixer associations to plants and soils. Ecosystems 19, 1491–1504.

Russell, N.J., 2007. Psychrophiles: membrane adaptations. In: Physiology and Biochemistry of Extremophiles.pp. 155–164.

Russell, N., 2008. Membrane components and cold sensing. In: Margesin, R., Schinner, F., Marx, J.-C., Gerday, C. (Eds.), Psychrophiles: From Biodiversity to Biotechnology. Springer-Verlag, Berlin Heidelberg, pp. 177–190.

Schadel, C., Bader, M.K.F., Schuur, E.A.G., Biasi, C., Bracho, R., Capek, P., De Baets, S., Diakova, K., Ernakovich, J., Estop-Aragones, C., et al., 2016. Potential carbon emissions dominated by carbon dioxide from thawed permafrost soils. Nature Clim. Change 6, 950–953.

Schaefer, K., Jafarov, E., 2016. A parameterization of respiration in frozen soils based on substrate availability. Biogeosciences 13, 1991–2001.

Schmid, B., Klumpp, J., Raimann, E., Loessner, M.J., Stephan, R., Tasara, T., 2009. Role of cold shock proteins in growth of *Listeria monocytogenes* under cold and osmotic stress conditions. Appl. Environ. Microbiol. 75, 1621–1627.

Schostag, M., Stibal, M., Jacobsen, C.S., Bælum, J., Taş, N., Elberling, B., Jansson, J.K., Semenchuk, P., Priemé, A., 2015. Distinct summer and winter bacterial communities in the active layer of Svalbard permafrost revealed by DNA- and RNA-based analyses. Front. Microbiol. 6.

Schuur, E.A.G., Bockheim, J., Canadell, J.G., Euskirchen, E., Field, C.B., Goryachkin, S.V., Hagemann, S., Kuhry, P., Lafleur, P.M., Lee, H., et al., 2008. Vulnerability of permafrost carbon to climate change: implications for the global carbon cycle. Bioscience 58, 701–714.

Schuur, E.A.G., Vogel, J.G., Crummer, K.G., Lee, H., Sickman, J.O., Osterkamp, T.E., 2009. The effect of permafrost thaw on old carbon release and net carbon exchange from tundra. Nature 459, 556–559.

Shaver, G.R., Billings, W.D., Chapin, F.S., Giblin, A.E., Nadelhoffer, K.J., Oechel, W.C., Rastetter, E.B., 1992. Global change and the carbon balance of arctic ecosystems. Bioscience 42, 433–441.

Shaver, G.R., Giblin, A.E., Nadelhoffer, K.J., Thieler, K.K., Downs, M.R., Laundre, J.A., Rastetter, E.B., 2006. Carbon turnover in Alaskan tundra soils: effects of organic matter quality, temperature, moisture and fertilizer. J. Ecol. 94, 740–753.

Shcherbakova, V., Chuvilskaya, N., Rivkina, E., Demidov, N., Uchaeva, V., Suetin, S., Suzina, N., Gilichinsky, D., 2013. *Celerinatantimonas yamalensis* sp. nov., a cold-adapted diazotrophic bacterium from a cold permafrost brine. Int. J. Syst. Evol. Microbiol. 63, 4421–4427.

Siewert, M.B., Hugelius, G., Heim, B., Faucherre, S., 2016. Landscape controls and vertical variability of soil organic carbon storage in permafrost-affected soils of the Lena River Delta. Catena 147, 725–741.

Srimathi, S., Jayaraman, G., Feller, G., Danielsson, B., Narayanan, P.R., 2007. Intrinsic halotolerance of the psychrophilic α-amylase from *Pseudoalteromonas haloplanktis*. Extremophiles 11, 505–515.

Steven, B., Leveille, R., Pollard, W.H., Whyte, L.G., 2006. Microbial ecology and biodiversity in permafrost. Extremophiles 10, 259–267.

Stevenson, A., Burkhardt, J., Cockell, C.S., Cray, J.A., Dijksterhuis, J., Fox-Powell, M., Kee, T.P., Kminek, G., McGenity, T.J., Timmis, K.N., et al., 2015. Multiplication of microbes below 0.690 water activity: implications for terrestrial and extraterrestrial life. Environ. Microbiol. 17, 257–277.

Strauss, J., Schirrmeister, L., Mangelsdorf, K., Eichhorn, L., Wetterich, S., Herzschuh, U., 2015. Organic-matter quality of deep permafrost carbon—a study from Arctic Siberia. Biogeosciences 12, 2227–2245.

Street, L.E., Dean, J.F., Billett, M.F., Baxter, R., Dinsmore, K.J., Lessels, J.S., Subke, J.A., Tetzlaff, D., Wookey, P.A., 2016. Redox dynamics in the active layer of an Arctic headwater catchment; examining the potential for transfer of dissolved methane from soils to stream water. J. Geophys. Res. Biogeosci. 121, 2776–2792.

Streletskiy, D.A., Sherstiukov, A.B., Frauenfeld, O.W., Nelson, F.E., 2015. Changes in the 1963–2013 shallow ground thermal regime in Russian permafrost regions. Environ. Res. Lett. 10, 10.

Stres, B., Philippot, L., Faganeli, J., Tiedje, J.M., 2010. Frequent freeze-thaw cycles yield diminished yet resistant and responsive microbial communities in two temperate soils: a laboratory experiment. FEMS Microbiol. Ecol. 74, 323–335.

Sturtevant, C.S., Oechel, W.C., 2013. Spatial variation in landscape-level CO_2 and CH_4 fluxes from arctic coastal tundra: influence from vegetation, wetness, and the thaw lake cycle. Glob. Chang. Biol. 19, 2853–2866.

Stutter, M.I., Billett, M.F., 2003. Biogeochemical controls on streamwater and soil solution chemistry in a High Arctic environment. Geoderma 113, 127–146.

Suetin, S.V., Shcherbakova, V.A., Chuvilskaya, N.A., Rivkina, E.M., Suzina, N.E., Lysenko, A.M., Gilichinsky, D.A., 2009. *Clostridium tagluense* sp. nov., a psychrotolerant, anaerobic, spore-forming bacterium from permafrost. Int. J. Syst. Evol. Microbiol. 59, 1421–1426.

Tarnocai, C., Canadell, J.G., Schuur, E.A.G., Kuhry, P., Mazhitova, G., Zimov, S., 2009. Soil organic carbon pools in the northern circumpolar permafrost region. Glob. Biogeochem. Cycles 23, 11.

Turner, B.L., Baxter, R., Mahieu, N., Sjogersten, S., Whitton, B.A., 2004. Phosphorus compounds in subarctic Fennoscandian soils at the mountain birch, (*Betula pubescens*)—tundra ecotone. Soil Biol. Biochem. 36, 815–823.

Vincent, A.G., Sundqvist, M.K., Wardle, D.A., Giesler, R., 2014. Bioavailable soil phosphorus decreases with increasing elevation in a subarctic tundra landscape. PLoS ONE. 9.

Vishnivetskaya, T., Kathariou, S., McGrath, J., Gilichinsky, D., Tiedje, J.M., 2000. Low-temperature recovery strategies for the isolation of bacteria from ancient permafrost sediments. Extremophiles 4, 165–173.

Vishnivetskaya, T.A., Petrova, M.A., Urbance, J., Ponder, M., Moyer, C.L., Gilichinsky, D.A., Tiedje, J.M., 2006. Bacterial community in ancient siberian permafrost as characterized by culture and culture-independent methods. Astrobiology 6, 400–414.

Vorobyova, E., Soina, V., Gorlenko, M., Minkovskaya, N., Zalinova, N., Mamukelashvili, A., Gilichinsky, D., Rivkina, E., Vishnivetskaya, T., 1997. The deep cold biosphere: facts and hypothesis. FEMS Microbiol. Rev. 20, 277–290.

Wagner, D., Schirmack, J., Ganzert, L., Morozova, D., Mangelsdorf, K., 2013. *Methanosarcina soligelidi* sp. nov., a desiccation- and freeze-thaw-resistant methanogenic archaeon from a Siberian permafrost-affected soil. Int. J. Syst. Evol. Microbiol. 63, 2986–2991.

Waldrop, M.P., Wickland, K.P., White, R., Berhe, A.A., Harden, J.W., Romanovsky, V.E., 2010. Molecular investigations into a globally important carbon pool: permafrost-protected carbon in Alaskan soils. Glob. Chang. Biol. 16, 2543–2554.

Walker, D.A., 2000. Hierarchical subdivision of Arctic tundra based on vegetation response to climate, parent material and topography. Glob. Chang. Biol. 6, 19–34.

Welsh, D.T., 2000. Ecological significance of compatible solute accumulation by microorganisms: from single cells to global climate. FEMS Microbiol. Rev. 24, 263–290.

Whittinghill, K.A., Hobbie, S.E., 2012. Effects of pH and calcium on soil organic matter dynamics in Alaskan tundra. Biogeochemistry 111, 569–581.

Wilhelm, R.C., Niederberger, T.D., Greer, C., Whyte, L.G., 2011. Microbial diversity of active layer and permafrost in an acidic wetland from the Canadian High Arctic. Can. J. Microbiol. 57, 303–315.

Wu, G., Wei, J., Deng, H.B., Zhao, J.Z., 2006. Nutrient cycling in an alpine tundra ecosystem on Changbai mountain, northeast China. Appl. Soil Ecol. 32, 199–209.

Yergeau, E., Hogues, H., Whyte, L.G., Greer, C.W., 2010. The functional potential of high Arctic permafrost revealed by metagenomic sequencing, qPCR and microarray analyses. ISME J. 4, 1206–1214.

Zhang, T., Barry, R.G., Knowles, K., Ling, F., Armstrong, R.L., 2003. Distribution of Seasonally and Perennially Frozen Ground in the Northern Hemisphere. A a Balkema Publishers, Leiden.

Geothermal steam vents of Hawai'i

Richard L. Weiss Bizzoco, Scott T. Kelley

Department of Biology, San Diego State University, San Diego, CA, United States

Chapter outline

1 Introduction

The island of Hawai'i lies in the mid-Pacific Ocean and with the Kilauea summit presents a highly active volcanic site. The island has a scarcity of flowing hot springs and bubbling pools commonly found in terrestrial hydrothermal fields (Ingebritsen and Scholl, 1993). Instead, there is an abundance of both mild ($\sim$40–60°C) and extreme (60–92°C) steam vent habitats. Consistent and abundant rainfall on the island, coupled with porous and sieve-like lava allows the groundwater recharge to descend to either shallow or greater depths and percolate through to the prominent magmatic heat supply. Whether shallow or deep, the contact between descending groundwater and rising heat results in a nearly continuous upward flow of steam that travels through porous passageways and forms numerous steam caves and vents ultimately escaping by horizontal or vertical cave/vent outlets.

Model Ecosystems in Extreme Environments. https://doi.org/10.1016/B978-0-12-812742-1.00002-7

Terrestrial volcanic areas have a number of hydrothermal features such as fumaroles, seeps, flowing springs (Boyd et al., 2007), and hot pools or terraces of various kinds (Bonheyo et al., 2005). Hawai'i has essentially only one type of hydrothermal feature, the steam cave/vent. Fumarolic steam vapors rising from caves or vents have been largely bypassed by investigators, yet steam vents remain as the dominant hydrothermal feature on Hawai'i and potentially provide a pathway to discover new information about microorganisms in extreme environments. Here, we describe organisms found in near neutral steam caves, identify a previously unknown type of steam cave, summarize our findings for unknown steam cave organisms, and describe key aspects of a broader survey of Hawaiian microorganisms (Wall et al., 2015).

Geothermal steam content is influenced by contact with magmatic gases such as SO_2 and H_2S as well as the ongoing extraction of lava by descending rainwater and heated steam rising to the surface. Prominent among these influences is the effect of volcanic gases that if present may result in forming highly acidic (pH 0–2), or if absent near neutral steam cave/vent deposits. Steam deposits are shaped by physical and chemical interactions as the rising steam contacts the cooler cave walls and ceilings and is altered, causing deposits to form by several processes. Sampling of steam cave surfaces was limited to the few mm of the steam deposit site where steam comes in direct contact with the cave ceiling or wall. The formation of steam deposits on lava can occur as follows: (1) If material in the steam contacts a cooler surface, water forms and evaporates, leaving behind dissolved salts including nitrates, sulfates, or others in the form of solid *evaporites*; (2) iron in the form of Fe(II) rises as a soluble ion in the heated steam, as it cools at the surface and changes its form, it deposits as Fe(III) on the cave ceiling forming *precipitates* of iron oxides, hydroxides, oxyhydroxides, or complex iron minerals; (3) *particulates* carried by steam from subsurface fissures, passageways, or crevices rise to the surface and become trapped or bound selectively within the steam deposit matrix; (4) magmatic gases such as SO_2 and H_2S are highly soluble at high pressure and temperature below ground. Rising to the surface with decreasing solubility and atmospheric oxygen, *sublimates* form as the gases oxidize forming elemental sulfur crystals; (5) *sedimentation* is a process that occurs in flowing springs. Material removed by erosion is carried downstream and deposits by settling. This term is applied to flowing hot springs, but the process does not apply to steam deposits. We correctly apply the term "steam deposit" to the first four processes before, but not to sedimentation, a process commonly used to describe movement and settling in flowing springs or bubbling pools.

When TD Brock first started work on thermal microbiology in Yellowstone National Park in the mid-to-late 1960s, one approach used was slide immersions in thermal springs. The result was the appearance of small unknown dark dot-like spheres on the slides at high temperatures (>70°C). Since their properties were unknown, cultures were established and they were still regarded as unusual. Electron microscopic studies recognized that these were not ordinary microbes with typical layered cell walls (Brock et al., 1972). Work on hot springs at that time concentrated on flowing springs and bubbling pools with temperature and chemical outflow gradients. Fumaroles were not part of these early investigations, despite their overwhelming abundance, for example, at Roaring Mountain there were hundreds of fumaroles and only one flowing hot spring. Fumaroles consisted only of steam and reached temperatures over 100°C. Since they appeared to lack viable cells (Brock, 1978), they were not investigated in these early years. Despite their abundance, there are still few diversity studies of steam vents (Mayhew et al., 2007; Ellis et al., 2008; Benson et al., 2011; Medrano-Santillana et al., 2017), most studies concentrated on the soils surrounding steam vents (Stott et al., 2008; Costello et al., 2009; Soo et al., 2009). Other investigations examined dispersal by steam rising from the surface of thermal pools and found sequences identifying eukaryotes, bacteria, and viruses (Bonheyo et al., 2005; Snyder et al., 2007). Studies in Hawai'i examined the diversity of eukaryotic organisms in milder steam vents (Ackerman et al., 2007) or bacteria associated with lava deposit age (Gomez-Alvarez et al., 2007). Bacteria were examined for diversity studies, rather than Archaea due to technical difficulties of isolating amplifiable DNA from steam vent samples (Herrera and Cockell, 2007; Benson et al., 2011; Medrano-Santillana et al., 2017). Even given the wide variety of DNA isolation kits and methods (Henneberger et al., 2006) this remains as a continuing problem with volcanic samples.

2 Steam cave and vent sites

Fumaroles or steam vents are found as an abundant geothermal feature on the Big Island of Hawai'i. As an example, hundreds of fumaroles are seen in the Kilauea area, but only one thermal pool has been found (Waite, 2007). Although they represent one of the least studied thermal features, they are the only terrestrial hydrothermal feature readily available for investigation on Hawai'i. Although once considered to lack living microbes due to their high temperatures, the steam-filled caves and vents of Hawai'i have been found to contain a wide diversity of microorganisms (Wall et al., 2015). Steam forms from meteoric waters that descend into the subsurface and meet upward convection of heat. The steam quickly forms, rising vertically to the surface through crevices and fissures in the lava. As it nears the surface, it fills horizontal or vertical caves and exits the passageway in a continuous or burst-like form, depending on the flow rate. Although some caves have high burst discharge, the steam can be considered to be an artesian flow. Physically steam caves can be either shallow, 1 m or less, or deep, several meters, with narrow or wide openings. The characteristics of the caves we selected for collection were developed with the assistance of the Hawai'i Volcanoes National Park cultural liaison, Keola Awong. The individual sites as per our collection agreement with the National Park are coded sampling sites.

3 Steam cave and vent sample collection

Sites were selected at Hawai'i Volcanoes National Park for their chemical and physical properties. Locations were often widely distant from each other, but on some occasions when their physical properties were similar, harbored a common

Table 1 Physical properties of sites sampled

Location	pH	Temp (°C)	Type	Chemistry
Hawaii Volcanoes National Park				
Hawai'i (H1)	6.4	65.0	H	Nonsulfur cave
Hawai'i (H2)	6.9	68.0	V	Nonsulfur cave
Hawai'i (H3)	3.1	82.0	H	Sulfur cave
Hawai'i (H4)	3.1	82.0	H	Iron/sulfur cave
Hawai'i (H5)	5.0	68.0	H	Salt cave
Hawai'i (H6)	ND	66.0	V	Salt/sulfur cave
Yellowstone National Park[1]				
Roaring Mountain	4.8	93.0	H	Nonsulfur cave
Kamchatka, Russia[2]				
Mutnovsky Volcano	3.9	94.0	V	Sulfur vent

H, horizontal; V, vertical; 1, 2, included for comparison.

organism, although the overall diversity of two sites still might vary, depending on their site chemistry. Table 1 presents the properties of the steam caves and vents sampled. We collected both steam and steam deposits from the sites sampled. Sites from Yellowstone, WY and Kamchatka, Russia are included for comparison.

Hawaiian sites have sulfur caves, iron/sulfur caves, salt/sulfur, and nonsulfur caves. The nonsulfur caves are simple steam caves/vents or salt caves. The salt caves result from leaching of lava by descending rainwater and rising heated steam followed by deposition of leached material by evaporation, yielding evaporites.

A nonsulfur cave is presented in Fig. 1A. Steam caves may appear to lack steam or have abundant steam depending on the flow out of the cave exit and the ambient temperature. Measurement of the temperature at the cave opening confirmed that there is a continuous steam flow in most of the steam caves. In addition to an abundant steam flow, site selection was also dependent on being able to access the interior of a cave for purposes of collecting aseptically. The caves selected had to have suitable openings and access for our collecting extension poles carrying our sterile collection tubes. These were essential features in addition to the cave characteristics of temperature and pH, presented in Table 1. Our goal with sampling was to complete an aseptic sampling. For that, we needed, in most cases, to be able to visualize where and how our extension poles were contacting surfaces within the steam caves. Fig. 1 shows a steam cave where no visual ability to see the sampling site exists. In this case, we had to simply use a tactile approach once we were able to gain entry with the extension pole-tube sampling device. The contact between the sampling tube and the cave ceiling was guided by the cave topography and surface characteristics. If the ceiling material was soft, visual contact was required. We were interested in collection at the contact site between steam and the deposit. So, shallow 2–5 mm sampling depth was our goal with collections. With hard lava surfaces, the lava limited our collection to this range. With soft surfaces as in salt deposits visual contact was

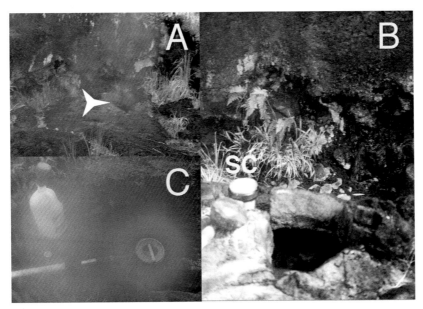

FIG. 1

(A) H1 steam cave shown with steam-filled air rising up and to the left. Three sided polygon marks steam cave opening at right, with rising steam. (B) Enlargement of H1 showing cave opening indicated in Fig. 1A and steam condenser to left (SC) capturing rising steam. (C) H2 site with steam condenser collection from continuous rising steam flow. Both sites H1 and H2 are nonsulfur steam caves.

essential during sample removal to keep sampling within our collection depth limits. As shown in Fig. 1A and B, the opening of the steam cave was quite narrow, with a length of about 30–35 cm and the cave was horizontal, limiting access deep inside the cave even with a 1–2 m long sampling pole. Still, by maneuvering in front of the opening at depth and collapsing the pole, we sampled 1.5 M inside the cave. After sampling had been completed, the tubes were closed and immediately chilled if they were to be used for DNA isolation. Samples for culture remained at ambient temperature, usually 25°C or less. For chemical characterization of our steam deposits, several different methods were utilized. We collected steam as shown in Fig. 1B and C using our portable steam condenser. We extracted deposits with nanopure water (Barnstead) to obtain soluble nutrients present in the deposits. We also carried out an HCl extraction, needed for some nutrients. We filtered (0.2 μm) both the collected steam and the deposit extracts and analyzed the samples by inductively coupled plasma optical emission spectroscopy (ICP, Table 2). We also analyzed the deposit samples by ion chromatography (IC, Table 3). The sulfur sites all obtain the sulfur from deeper faults that reach the magma below. Sulfur deposits form from hydrogen sulfide that oxidizes at the surface and deposits as sulfur sublimates. Sulfur dioxide, also present dissolves in the ground water, making hydrogen sulfide the main contributor to elemental sulfur in the sulfur caves in Hawai'i. When iron deposits on

Table 2 Steam vent steam chemistry based on ICP and nutrient analysis

Analyte (mg L^{-1})	RMNS° CST	RMNS° C	KRS° VST	H1NS° C	H2NS° C	H3S°C	H3S° CST
Na	0.666	2.421	1.43	3.454	3.463	3.360	1.57
Ca	0.00	0.596	0.452	2.285	3.592	4.205	0.789
Al	0.00	4.432	0.423	1.024	1.082	4.052	0.281
Fe (total)	0.00	3.084	0.163	0.975	0.998	2.755	0.335
Si	0.0676	19.311	8.20	18.579	29.222	21.326	0.482
B	0.00	2.332	0.369	0.697	0.700	0.705	0.505
K	0.00	0.371	0.631	2.864	2.519	2.510	1.22
Mg	0.00	0.0245	0.330	0.119	1.434	1.892	0.0451
Zn	0.127	0.0598	1.09	0.0147	0.0195	0.315	0.0231
Mn	0.00	0.00	0.00	0.00604	0.00646	0.0532	0.0189
Mo	0.0270	0.00	0.00	0.00	0.00	0.00	0.00
Se	0.00	0.0171	0.0184	0.0289	0.0220	0.0242	0.00
Ni	0.00	0.0173	0.0122	0.00	0.00	0.00	0.00573
Pb	0.00165	0.00	0.00	0.00	0.00	0.00	0.00
Cr	0.00535	0.00	0.00	0.00	0.00	0.00	0.00557
Cd	0.00	0.0229	0.0401	0.00164	0.00215	0.00199	0.0137
Cu	0.00	0.498	0.00	0.00	0.00	0.809	0.00
Hg	0.00	0.00	0.0183	0.00	0.00	0.00	0.00
As (total)	0.00059	0.0109	0.0128	0.0242	0.0266	0.0435	0.00
S	5.181	50.763	1.09	0.0312	0.00	14.474	0.146
Sr	0.0311	0.00459	0.00435	0.0832	0.00858	0.00	0.00
NO$_2$/NO$_3$, N-NO$_3$ (μM)	3.007.	1.443	3.471	0.00	0.00	0.000	2.143
NH$_4$,N-NH$_4$ (μM)	0.011	18.391	0.529	19.550	79.023	3.650	0.244
PO$_4$,P-PO$_4$ (μM)	0.067	5.593	0.00	0.00	27.372	0.00	0.0910
SO4 (μM)	781.3	355.0	7.5	0.00	0.00	185.00	30.0
Conductivity (μScm-1)	47.6	1420.0	22.3	13.9	29.3	257.0	8.6

RMNS°CST, roaring mountain nonsulfur cave steam; RMNS°C, roaring mountain nonsulfur cave; KRS° VST, Kamchatka Russia sulfur vent steam; H1NS°C, Hawai'i 1 nonsulfur cave; H2NS°C, *Hawai'i 2 nonsulfur cave;* H3S°C, Hawai'i 3 sulfur cave;H3S°CST, *Hawai'i 3 sulfur cave steam.*

sulfur, the iron is kept in soluble form Fe(II) by acidic conditions. As steam vapors and air mix near the opening of steam caves, iron begins to oxidize to Fe(III) and deposits form on the sulfur as reddish-brown iron oxides or hydroxides. The abundant steam makes these sites nutrient-rich habitats for high-temperature sulfur-oxidizing organisms. Once we recognized that organisms were present in flowing steam, we considered that they might be present in all steam deposits that accumulated along the cave ceilings and walls. We began to examine what types of caves might be found on the island of Hawai'i with the idea that in addition to sulfur, iron, and nonsulfur caves, other kinds of chemical steam saturated habitats might be found. Our search for niches caused us to travel over new volcanic areas searching

Table 3 Steam vent steam chemistry based on ion chromatography

Conc (µM)	RMNS °C	KRS° VST	H2NS °C	H3S°C	H6SCST	HIS °V	HIS° VST
Acetate	346.1	34.1	21.4	469.8	1.2	0.5	4.6
Bromide	0.0	0.4	0.2	8.9	0.0	0.1	0.0
Chloride	7.7	16.9	2.0	514.0	9.1	5.3	112.8
Citrate	0.8	0.1	0.0	0.0	0.0	0.0	0.0
Fluoride	0.3	0.2	0.2	7.3	0.8	1.3	0.7
Formate	114.4	1.1	25.1	679.5	1.7	0.9	0.7
Lactate	0.0	0.0	0.9	63.6	0.0	0.5	0.0
Nitrate	64.5	124.6	376.3	7207.2	0.0	76.2	30.8
Nitrite	0.6	0.0	0.0	0.0	0.0	0.5	0.0
Oxalate	14.5	15.7	5.8	142.5	8.1	2.7	28.0
Phosphate	1.8	1.4	2.2	74.5	0.0	0.6	0.6
Sulfate	9.3	44.0	0.6	394.6	2.5	4.0	639.5

RMNS°C, *roaring mountain nonsulfur cave;* KRS°VST, *Kamchatka Russia sulfur vent steam;* H2NS°C, *Hawai'i 2 nonsulfur cave;* H3S°C, *Hawai'i 3 sulfur cave;* H6SCST, *Hawai'i 6 salt cave steam;* HIS°V, *Hawai'i sulfur vent, no code number;* HIS°VST, *Hawai'i sulfur vent steam, no code number.*

for something unknown. We began to notice siliceous sinter and salt-like deposits and felt they were of significant interest. As we viewed cave interiors, our views initially were obscured by thick clouds of flowing steam, not necessarily from those caves with the most extreme temperatures. As shown in Fig. 2, such caves were unusual in that they did represent a new steam cave type for future collections.

3.1 Steam collection

Our steam collector (Fig. 1B and C) worked well using nonsterile water at ambient temperature, ($\sim$23°C) to supply the condenser. As steam condensed, the temperature of the upper condenser increased. The appearance of bubbles in the water signaled when to replace the condenser water. The rate of condensation was slightly less than 1 mL min^{-1} in the hottest Hawaiian vents (82°C) and the rate varied with the temperature and flow rate of steam. Steam water was collected for water chemistry and culture. Cell counts using DAPI staining yielded around 600 cells mL^{-1}. Cells observed were either spherical, irregular, or rod shaped and no thin filaments were seen in steam collected samples. Control samples contained no microbial cells. We obtained clones and sequences from high-temperature steam collected at site H3. However, it was necessary to use a GenomiPhi kit to amplify the DNA collected from the steam samples. We obtained only halophiles from the steam collected at H3 (Ellis et al., 2008). In our subsequent collections we investigated steam deposits and used the collected samples to isolate and purify DNA or to grow cultures. An important aspect of steam collection for chemistry and microbiology is that the collector was easy to sterilize both in the laboratory and once we were in the field so that it could be

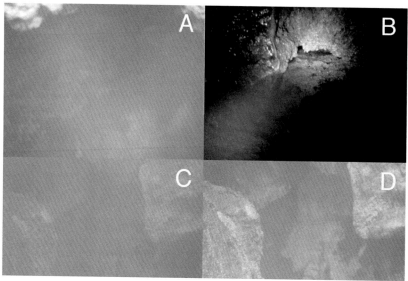

FIG. 2

Salt caves. H5 (A) Salt cave opening surrounded by steam. (B) Visible only at night time with LED flashlight photo, salt cave deposits on floor and ceiling lie several meters deep within cave. H6 (C) Salt/sulfur cave experiences continuous bursts of steam obscuring all views. (D) Salt/sulfur cave deposit seen deep inside steam-filled cave with yellow sulfur surrounded on either side by white salt deposits.

used at multiple sites with presterilized condensers. It was also important that the condenser was small and easily transported over long distances by air travel.

Unique weather patterns and high volcanic activity make Hawai'i a place where unusual conditions can be established; these appear to persist over time and provide habitats with singular physicochemical conditions that determined temperature, pH, and chemical attributes. As an example, the salt caves shown in Fig. 2 represent habitats where the temperature is high deep within the steam cave and there is a salt bed 5–10 cm deep of a white crystalline deposit (Fig. 2A and B) The deposit occurs at a depth of ∼5 m within the cave and can only be seen at night due to the nearly continuous flow of hot steam and moderately high temperature. In the second cave type, salt/sulfur, the deposit lies deep within the cave and the salt and sulfur deposits form closely together (Fig. 2C and D). To our knowledge, other reports of this type of steam cave habitat found outside of Hawai'i have not appeared.

3.2 Steam deposit collection

Steam deposits by their nature are more difficult to collect than steam. First, the deposit material lies within an enclosed cave or vent and so is more difficult to access. Second, the collection site may be difficult to identify, or in some cases as presented before, the site cannot be seen at all. We carried out aseptic collections using a shallow sampling, few mm at most. With a pole and sterile 50 mL collection tube angled toward the investigator, the deposit was scraped from the ceiling by the upper edge of the open tube so the sample fell into the tube and was capped immediately after collection. The deposits were attached to sample stubs with a carbon adhesive tab and analyzed with an X-Max 50 mm^2 energy dispersive X-ray detector and a Quanta 450 FEI-SEM, using Oxford Inca software. The spectra of three different steam caves shown in Fig. 3A–C, include (A) Hawai'i H1 nonsulfur cave, (B) Hawai'i H2 nonsulfur cave, and (C) Hawai'i H4 iron/sulfur cave. H1 and H2 represent higher temperature 65–68°C near neutral pH caves; both nonsulfur caves are shown in Fig. 1B and C, respectively; H4 is a more extreme 82°C acidic site. The deposits in all three sites have supplied nutrients essential for the growth of microorganisms in the steam deposits sites.

In the steam deposit spectra shown (Fig. 3A–C) Hawai'i nonsulfur cave H1 (A) showed high peaks for aluminum (Al), silicon (Si), and iron (Fe). A smaller sulfur (S) peak was always present along with peaks for oxygen (O), titanium (Ti) and several important cofactors, potassium (K), sodium (Na), magnesium (Mg), and calcium (Ca) were seen as small peaks. The small sodium (Na) peak appears just before magnesium (Mg) in Fig. 3A and B. Titanium (Ti) was found in most spectra, a role in metabolism is unknown and it appears with silicon in the Ka'u Desert, HI as a surface coat on rocks (Chemtob et al., 2006). The nonsulfur cave deposit Hawai'i 2 (H2) spectrum (Fig. 3B) was similar to (Fig. 3A) in having significant peaks for aluminum (Al), silicon (Si), and iron (Fe). A sulfur (S) peak was not detected, suggesting no access was available to deep magmatic gases. The iron (Fe) peak was slightly greater in this spectrum (Fig. 3B) and in other Hawai'i 2 H2 spectra (data not shown), compared to the iron peak of Hawai'i 1 H1 (A). Part of the manganese (Mn) signal, K_β, coincides with the iron peak and may contribute to the iron peak size. Oxygen (O) was also present in this spectrum (Fig. 3B). In addition, Hawai'i 2 H2 lacked a potassium (K) peak and the sodium (Na) peak was minor. The magnesium (Mg) and calcium (Ca) peaks were slightly greater than those of Hawai'i 1 H1 (A). Hawai'i 2 H2 (Fig. 3B) had a minor titanium (Ti) peak, approximately half that of the titanium (Ti) peak of the Hawai'i 1 H1 nonsulfur cave sample (Fig. 3A). Manganese (Mn) was also detected as a modest peak of Hawai'i 2 H2, while no manganese was present in Hawai'i H1 (Fig. 3A). The significance of manganese in this sample is unknown.

Hawai'i 4 H4 iron/sulfur cave (Fig. 3C) had only modest peaks for aluminum (Al) and silicon (Si), while prominent peaks of aluminum (Al) and silicon (Si) were seen in spectra for Hawai'i H1 and H2 nonsulfur caves (Fig. 3A) and (Fig. 3B), respectively. Similarly, the iron (Fe) peak was modest with defined K and L lines, and peak heights appeared to be less than the peak heights seen in Hawai'i 1 H1 (Fig. 3A) and Hawai'i 2 H2 (Fig. 3B), despite the observation that iron/sulfur sample H4 was reddish in color, and suggested the presence of iron oxides, hydroxides, or oxyhydroxides. The most prominent peak, sulfur (S), dominated the sample and was high in Hawai'i 4 H4 compared to the sulfur peak of Hawai'i 1 H1 (Fig. 3A). No sulfur peak appeared in Hawai'i 2 H2 (Fig. 3B). The titanium (Ti) peak appeared minor, but

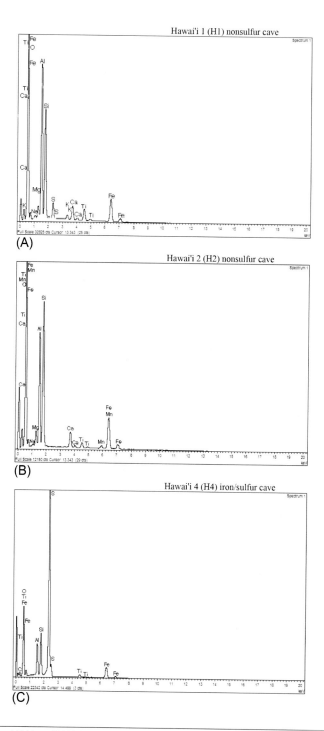

FIG. 3

X-ray spectra: (A) Hawai'i H1 Nonsulfur; (B) Hawai'i H2 Nonsulfur; (C) Hawai'i H4 iron/sulfur; S° sublimate-iron over sulfur. Baseline scale in Fig. 3A–C is from 0 to 20 keV.

FIG. 4

SEM images of scanned steam deposit samples used to generate X-ray spectra above. The letters indicate the spectrum source sample. SEM (A) Hawai'i 1 H1 nonsulfur cave, a band of particulate matter angled left in middle represents lava from cave-ceiling scrapings; SEM (B) Hawai'i 2 H2 nonsulfur cave, small and large particles are from scrapings of lava cave-ledge underside; SEM (C) Hawai'i 4 H4 iron/sulfur cave. Rhomboidal sulfur crystals seen at top middle and top right appear bright, a result of charging. SEM (A-C) Small particles were evident, but no cells, walls, or appendages seen in Fig. 4A–C. (A–C) Bar, 0.5 mm.

detectable compared to those of Hawai'i 1 and 2 (Fig. 3A and B). A minor carbon (Fig. 3C) peak was present in Hawai'i 4 H4 (iron/sulfur), while none was observed in Hawai'i H1 (Fig. 3A) or H2 (Fig. 3B). This peak suggested the possibility of cellular material. We examined several SEM images associated with spectra from each site (Fig. 3A–C). Three representative SEM images are shown (Fig. 4). Each spectrum has an SEM image with the same letter; the counterpart for spectrum Fig. 3A is Fig. 4A. No cells, walls, or appendages were seen. Despite the negative SEM results, the carbon peak (Fig. 3C) may represent variable amounts of entrapped cells or walls, an unknown that remains to be controlled.

3.3 Steam and steam deposit collection: Control methods

Steam methods. Steam control samples included artificially generated steam from filter sterilized, autoclaved nanopure water (Barnstead). The control collection, carried out using our portable steam condenser, was completed adjacent to the steam vent or cave or in close proximity. Following collection, control samples were analyzed along with the steam cave/vent samples by PCR, filtration of a 10–15 mL sample on black 13 mm Millipore membranes (0.22 μm pore size) followed by DAPI microscopy, and by culture of thermophiles 50°C, 70°C, and 85°C. The controls were negative.

　　Steam deposit methods. The steam cave nonsulfur controls H1 and H2 samplings were from solid lava cave ceiling surfaces. While it was not possible to go below the steam deposit sample matrix, the controls for the lava surface were done on a nearby nonsteam ambient temperature lava cave or vent in identical fashion with scraping and tube collection used for steam cave/vent samples. The remaining samples were all visually accessible shallow samplings of thick deposits and did not include extraneous materials within the sampled steam deposit site. Control samples for steam deposits were evaluated by PCR, DAPI-microscopy, and thermophilic culture as described before for steam, and proved to be negative.

4 DNA isolation and phylogeny

Diversity studies on Hawaiian fumaroles lack a single effective method to isolate DNA from fumarole deposit samples, although several methods are available (Henneberger et al., 2006; Herrera and Cockell, 2007). This represents a current barrier to fully effective diversity studies that we are approaching in our work. More work is needed in this area though our neutralization-protein approach (Benson et al., 2011) has proven useful in this regard. Table 4 presents sequences that identified Archaea in 4 Hawaiian steam cave sites (Benson et al., 2011). Fumaroles exposed to sunlight such as H1 and H5 were surrounded by a ring of green (Figs. 1B and 2B) and sequences obtained showed these sites supported diverse populations of Cyanobacteria and Chloroflexi (green nonsulfur bacteria). Dark steam cave interiors were devoid of these photosynthetic bacteria (Wall et al., 2015). Clones from 3 Hawaiian nonsulfur cave interiors H1, H2, and a noncoded fumarole site clustered with group 1 Crenarchaeota and returned sequences closely related to a cultured thermophilic ammonia oxidizing archaeon, "*Candidatus Nitrosocaldus yellowstonii*" (de la Torre et al., 2008) and "*Candidatus Nitrosocaldus* sp." (AB201308). We note that where sequences related to ammonia oxidizing archaea (AOA) were found, the Hawaiian cave sites had high concentrations of ammonia (Table 2, H1, H2), neutral or near neutral pH, and high temperatures between 65° C and 68°C. Marine AOA are widespread and important contributors to the global nitrogen cycle (Urakawa et al., 2010). An increasing number of sequences with a characteristic set of marker genes led to the proposal of a new phylum, *Thaumarchaeota* to include marine AOA and closely related organisms (Brochier-Armanet et al., 2008).

Table 4 Archaeal sequences from steam vent clones with top BLAST matches[a]

Sample site	Accession	%ID	Organism
H1	DQ791908	99	Uncultured archaeon
	AB213053	95	Uncultured archaeon gene
H2	DQ791908	99	Uncultured archaeon
H3	AY907890	98	*Sulfolobus* sp.
	CP001401	97–98	*Sulfolobus islandicus*
H4	AY907890	96–99	*Sulfolobus* sp.
	CP001399	98	*Sulfolobus islandicus*
	DQ179001	99	Uncultured archaeon

[a]Archaeal clones related to thermophilic ammonia oxidizing archaea were recovered from H1, H2 and a noncoded nonsulfur cave (Benson et al., 2011; Wall et al., 2015).

5 **Culture isolation**

Our culture isolations from steam deposits met with mixed success. Direct microscopic visualization of steam deposit samples using DAPI staining revealed the presence of several types of cells in nonsulfur caves (Fig. 5). Fig. 5A and B shows cells from site H1, that seem similar, to "*Candidatus Nitrosocaldus yellowstonii*" isolated from a Yellowstone hot spring (de la Torre et al., 2008). H1 vent 2 phase contrast showed short chains of rods adjacent to oval cells with dark thick walls and a refractile appearance (Fig. 5C and D). Individual cells often can be recognized in steam deposit samples, but in other cases sample complexity obscured attached microorganisms (Fig. 6). For example, in Fig. 6A a large particulate silica-iron oxide deposit stained with DAPI reveals positively stained material resembling cells, but the sample matrix is complex, making unequivocal interpretation difficult. Another complicating factor is that silicon dioxide may exhibit pale blue fluorescence when illuminated with UV light used with DAPI examination of samples. Hawai'i H1 and H2 steam caves are both nonsulfur caves and steam deposits collected from both sites returned 16S rRNA gene sequences related to high-temperature ammonia-oxidizing archaea. However, images from direct visualization of steam deposit samples from H1 (Fig. 5A and B) and H2 (Fig. 6A and B) are distinctive and quite different from each other.

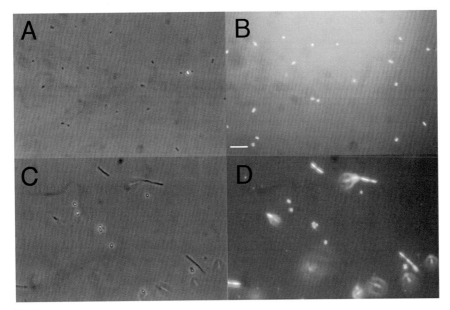

FIG. 5

Steam deposit samples from H1. (A, B) Vent 1 contains mainly small oval to small rod-shaped cells. (A) Phase contrast; (B) DAPI stained. (C, D) In nearby vent 2, short chains of rods were present along with oval refractile cells. (C) Phase contrast; (D) DAPI stained. (A–D) Bar, 5 μm.

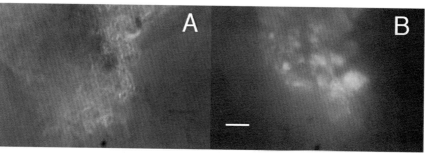

FIG. 6

H2 nonsulfur steam cave ledge deposits. Collections made several meters in the cave interior. Steam deposit collected from the protruding ledge bottom at point of steam-cave ceiling contact. Deposits formed reddish-clear iron-silica type matrix. (A) Phase contrast; (B) DAPI stained. (A, B) Bar, 10 μm.

Such differences might be related to site chemistry and steam flow, or sample collections and processing, all of which may influence appearances and visually mask site diversity.

When enrichments were carried out with sulfur cave deposits, no archaea were grown in culture (Benson et al., 2011). Archaea were only seen in culture at H1, a nonsulfur site. In contrast, bacterial cultures, both enrichments and isolated cultures were obtained at almost all sites examined, except H4, an iron-rich site. Salt caves represent an interesting unknown and unexplored steam cave habitat. We have enriched cultures from one of the salt caves identified in Table 1 as H5 and have seen rod-shaped organisms and very thin filaments ($\sim$0.2 × 5–10 μm) in cultures, but have not completed any identifications. Finding extremely thin filaments in salt caves was unexpected. Typically Archaea might be favored organisms, along with a variety of bacteria, some of which likely would be spore formers. In the case of very thin filaments (0.15–0.2 μm), they were not abundant and subculture on solid Gelrite or agar plates failed.

Cultures were established by enrichment in liquid medium at pH 4.5, 55°C for subsequent isolation on solid medium. Almost all sites were positive, growing bacteria in a range from 55°C to 85°C, usually on Gelrite plates or slants. Most sequences returned from isolated cultures aligned with Firmicutes, even when many sequences were obtained from strains with strikingly different pigmentations.

6 Environmental models

Models of the ammonia- and sulfur-oxidizing environments are presented below to provide an overview of the two processes as they occur in situ in their separate and distinct environments (Figs. 7 and 8).

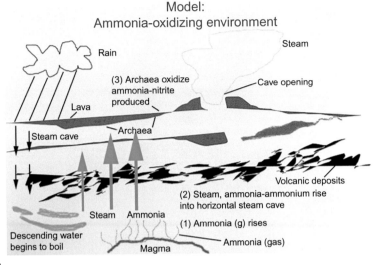

FIG. 7

Ammonia-oxidizing environment. Ammonia gases rise from underlying magma and travel upward through fissures and cracks (1). Ammonia is ionized and available as ammonium ion due to the favorable temperature and pH of the habitat. Ammonia-ammonium ions travel with rising steam, contact steam cave ceilings (2) and then pass horizontally over Archaea attached to ceiling steam deposits. A ready supply of ammonia-ammonium ion allows uptake and ammonia oxidation by ammonia-oxidizing archaea (3).

7 Conclusions

The abundance of fumaroles, their high microbial diversity, and worldwide distribution make this an important biological resource. The presence of organisms utilizing key metabolic nutrients deserves further investigation. If the Hawaiian ammonia-oxidizing archaea can be demonstrated by enrichment culture, amoA gene sequences, or other methods, then this demonstrates another new habitat for ammonia-oxidizing archaea and it seems to suggest an interesting parallel of ammonia oxidation in neutral/basic habitats and H_2S concentration and oxidation in acidic habitats. Our analysis so far has revealed novel archaeal lineages and nonsulfur steam deposits seem to be a rich source of unknown novel thermophilic organisms worthy of future investigation.

Acknowledgments

We thank Rhonda Loh, Chief of Natural Resources Management and Keola Awong, Cultural Specialist and Native Hawaiian Liaison, for facilitating our collections of fumaroles and supporting our research efforts in Hawai'i Volcanoes National Park. Jeff Sutton and Aaron

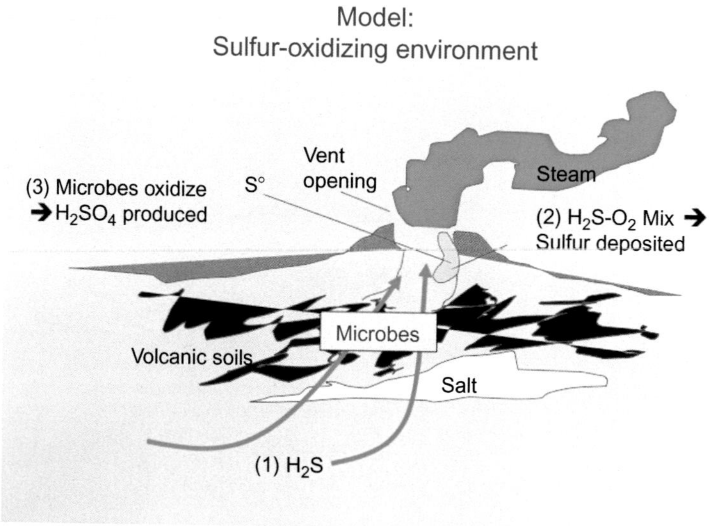

Model:
Sulfur-oxidizing environment

FIG. 8

Sulfur-oxidizing environment. Cracks and crevices descend deeply toward the magma, allowing hydrogen sulfide gases (1) to travel upward, eventually reaching the vent opening. There, H_2S mixes with oxygen, forming elemental sulfur deposits (2). Microbes, Archaea, attach to the sulfur crystals, oxidizing sulfur into sulfuric acid (3).

Pietruszka, US Geological Survey, offered guidance on the Geology/Geochemistry of our collection areas. We acknowledge the use in Table 1, the first four lines of data describing sites H1, H2, H3, H4, from Benson et al. (2011). Microbial diversity in nonsulfur, sulfur, and iron geothermal steam vents, by permission of Oxford University Press, Federation of European Microbiological Societies. Benno Spingler and Xzayla Zabiti provided ideas and participated in the design and construction of the steam collector. We acknowledge the contributions of Lisa Thurn who performed ICP experiments, David Lipson who carried out IC experiments and our graduate students, Dean Ellis, Courtney Benson, Kate Wall, and Jenny Cornell, and undergraduates, Christa Anderson and Wendy Gutierrez. Generous contributions of Schering-Plough Biopharma assisted this study. We appreciate the review comments for this manuscript.

References

Ackerman, C.A., Anderson, S., Anderson, C., 2007. Diversity of thermophilic microorganisms within Hawaiian fumaroles. Eos Trans. Am. Geophys. Union. 88.

Benson, C.A., Bizzoco, R.W., Lipson, D.A., Kelley, S.T., 2011. Microbial diversity in nonsulfur, sulfur and iron geothermal steam vents. FEMS Microbiol. Ecol. 76, 74–88.

Bonheyo, G.T., Frias-Lopez, J., Fouke, B.W., 2005. A test for airborne dispersal of thermophilic bacteria from hot springs. In: Inskeep, W.P., McDermott, T.R. (Eds.), Geothermal

Biology and Geochemistry in Yellowstone National Park. Montana State University Publications, Bozeman, MT, pp. 327–340.

Boyd, E.S., Jackson, R.A., Encarnacion, G., Zahn, J.A., Beard, T., Leavitt, W.D., et al., 2007. Isolation, characterization, and ecology of sulfur-respiring *Crenarchaea* inhabiting acid-sulfate-chloride-containing geothermal springs in Yellowstone National Park. Appl. Environ. Microbiol. 73, 6669–6677.

Brochier-Armanet, C., Boussau, B., Gribaldo, S., Forterre, P., 2008. Mesophilic Crenarchaeota: proposal for a third archaeal phylum, the Thaumarchaeota. Nat. Rev. Microbiol. 6, 245–252.

Brock, T.D., 1978. Thermophilic Microorganisms and Life at High Temperatures. Springer, New York.

Brock, T.D., Brock, K.M., Belly, R.T., Weiss, R.L., 1972. *Sulfolobus*: a new genus of sulfur-oxidizing bacteria living at low pH and high temperature. Arch. Mikrobiol. 84, 54–68.

Chemtob, S.M., Jolliff, B.L., Arvidson, R.E., 2006. Si- and Ti-rich surface coatings on Hawaiian basalt and implications for remote sensing on Mars. In: Lunar Planet. Sci. 37th Conference.

Costello, E.K., Halloy, S.R.P., Reed, S.C., Sowell, P., Schmidt, S.K., 2009. Fumarole-supported islands of biodiversity within a hyperarid, high-elevation landscape on Socompa Volcano, Puna de Atacama, Andes. Appl. Environ. Microbiol. 75, 735–747.

de la Torre, J.R., Walker, C.B., Ingalls, A.E., Könneke, M., Stahl, D.A., 2008. Cultivation of a thermophilic ammonia oxidizing archaeon synthesizing crenarchaeol. Environ. Microbiol. 10, 810–818.

Ellis, D.G., Bizzoco, R.W., Kelley, S.T., 2008. Halophilic *Archaea* determined from geothermal steam vent aerosols. Environ. Microbiol. 10, 1582–1590.

Gomez-Alvarez, V., King, G.M., Nusslein, K., 2007. Comparative bacterial diversity in recent Hawaiian volcanic deposits of different ages. FEMS Microbiol. Ecol. 60, 60–73.

Henneberger, R.M., Walter, M.R., Anitori, R.P., 2006. Extraction of DNA from acidic, hydrothermally modified volcanic soils. Environ. Chem. 3, 100–104.

Herrera, A., Cockell, C.S., 2007. Exploring microbial diversity in volcanic environments: a review of methods in DNA extraction. J. Microbiol. Meth. 70, 1–12.

Ingebritsen, S.E., Scholl, M.A., 1993. The hydrogeology of Kilauea volcano. Geothermics 22, 255–270.

Mayhew, L.E., Geist, D.J., Childers, S.E., Pierson, J.D., 2007. Microbial community comparisons as a function of the physical and geochemical conditions of Galápagos Island fumaroles. Geomicrobiol J. 24, 615–625.

Medrano-Santillana, M., Souza-Brito, E.M., Duran, R., Gutierrez-Corona, F., Reyna-López, G.E., 2017. Bacterial diversity in fumarole environments of the Paricutín volcano, Michoacán (Mexico). Extremophiles 21, 499–511.

Snyder, J.C., Wiedenheft, B., Lavin, M., Roberto, F.F., Spuhler, J., Ortmann, A.C., Douglas, T., Young, M., 2007. Virus movement maintains local virus population diversity. Proc. Natl. Acad. Sci. U. S. A. 104, 19102–19107.

Soo, R.M., Wood, S.A., Grzymski, J.J., McDonald, I.R., Cary, S.C., 2009. Microbial biodiversity of thermophilic communities in hot mineral soils of Tramway Ridge, Mount Erebus, Antarctica. Environ. Microbiol. 11, 715–728.

Stott, M.B., Crowe, M.A., Mountain, B.W., Smirnova, A.V., Hou, S., Alam, M., Dunfield, P.F., 2008. Isolation of novel bacteria, including a candidate division, from geothermal soils in New Zealand. Environ. Microbiol. 10, 2030–2041.

Urakawa, H., Martens-Habbena, W., Stahl, D.A., 2010. High abundance of ammonia-oxidizing *Archaea* in coastal waters, determined using a modified DNA extraction method. Appl. Environ. Microbiol. 76, 2129–2135.

Waite, M., 2007. Mosses of Hawai'i Volcanoes National Park. Technical Report 153. Pacific Cooperative Studies Unit, University of Hawai'i at Mānoa, p. 18.

Wall, K., Cornell, J., Bizzoco, R.W., Kelley, S.T., 2015. Biodiversity hot spot on a hot spot: novel extremophile diversity in Hawaiian fumaroles. Open Microbiol. J. 4, 267–281.

Solar salterns as model systems for the study of halophilic microorganisms in their natural environments

3

Aharon Oren

Department of Plant and Environmental Sciences, The Hebrew University of Jerusalem, The Edmond J. Safra Campus, Jerusalem, Israel

Chapter outline

1 Introduction

Saltern evaporation and crystallizer ponds are perfect field laboratories and model systems for the study of halophilic microorganisms: halotolerant algae and cyanobacteria, moderately halophilic heterotrophic bacteria, and extremely halophilic Archaea and Bacteria. Information obtained from research on salterns has yielded valuable information on the heterotrophic activities of Archaea and Bacteria based on the effects of specific inhibitors, the possible activity of halocins, the function of gas vesicles in *Haloquadratum*, and on the presence of different types of organisms based on lipids and pigments as specific biomarkers (Javor, 1983, 1989; Oren, 2002, 2005). While natural salt lakes are often subject to long-term changes in water level (Javor, 1989; Oren, 2002), salinity, and biological properties, the properties of the salterns are much more constant over time. In many systems salt production is a

Model Ecosystems in Extreme Environments. https://doi.org/10.1016/B978-0-12-812742-1.00003-9

seasonal activity, while other salterns are operated all year round where climatic conditions enable this. But from year to year each saltern system presents a well-reproducible gradient of salinity in which the salt concentration of each pond is kept more or less constant and develops its characteristic communities of halophilic and halotolerant organisms adapted to the local salinity. Salterns are also readily accessible so that the logistics of sampling are simple.

The biota of the salterns not only have a great scientific interest, but also have a direct impact—in some cases favorable, in other cases unfavorable—on the quality and the quantity of the salt produced (Javor, 2002). Although the nature of the "missing link" between saltworks biology and solar salt quality is still far from clear (Oren, 2010), an in-depth understanding of the biota of the salterns and the interactions between the different biological processes are of great interest to the salt-producing industry.

The number of published studies on the microorganisms inhabiting the brines and the sediments of solar saltern ponds is very large. For example, most new species of moderately and extremely halophilic Archaea and Bacteria described in the past decades were retrieved from salterns. Therefore this short review on salterns as model systems is mainly restricted to information obtained from two very well-studied systems: the salterns of Salt of the Earth Ltd. on the Red Sea coast of Israel and the Bras del Port salterns in Santa Pola near Alicante on the Mediterranean coast of Spain. Where relevant, this information is supplemented with data collected during studies of salterns at other geographic locations.

2 The community structure of prokaryote and eukaryote microorganisms in the salterns

Because of their high community densities and the (at least relatively) simply community structure of their biota, saltern systems worldwide have become preferred model systems for the elucidation of the microbial diversity at different salt concentrations. Such studies have included both cultivation-dependent and cultivation-independent techniques.

16S rRNA gene library-based cultivation-independent methods have been used in the Eilat salterns to investigate the community composition within the gypsum crust (Sørensen et al., 2005) (Fig. 1) and the bacterial community associated with brine shrimps (*Artemia* spp.) along the discontinuous salinity gradient. *Halomonas*- and *Salinivibrio*-derived sequences were abundantly found, and indications were obtained for the presence of a new phylum affiliated with *Chlorobium* and the *Bacteroidetes* groups associated with *Artemia* (Tkavc et al., 2011).

Classic morphology-based taxonomic methods were used to assess the changes in composition of the chlorophyll-containing phototrophic microbial communities (cyanobacteria, diatoms, green microalgae) along the salinity gradient in the Eilat salterns (Řeháková et al., 2009).

FIG. 1

A gypsum crust from the bottom of a saltern evaporation pond in Eilat. Salinity: ~20%.

In the framework of a cultivation-dependent study of yeast diversity in a variety of hypersaline environments, only one type of yeast was recovered from the Eilat salterns: *Trichosporon mucoides*, found at 14%–33% salt (Butinar et al., 2005).

In the past three decades the Alicante saltern system has been the best studied site for the elucidation of the prokaryote diversity in saltern systems. This is true for cultivation-dependent as well as for cultivation-independent approaches. Since the early cultivation-dependent survey of the site using a limited number of culture media, studies performed long before molecular cultivation-independent techniques became available (Rodriguez-Valera et al., 1985) we have learned much about the nature of the dominant organisms in the system. One of the two well-investigated isolates of the flat square archaeon *Haloquadratum walsbyi* was obtained from a Santa Pola crystallizer pond (Bolhuis et al., 2004; Burns et al., 2007). The Alicante saltern system was also included in a comprehensive survey of the extremely halophilic cultivable prokaryotes in eight Mediterranean, Atlantic, and Pacific solar salterns. Isolates retrieved from 32.8% and 34.4% salinity ponds yielded 23 operational taxonomic units, 54.5% of which belonged to the Bacteria (Viver et al., 2015).

The Santa Pola, Alicante saltern was the first hypersaline environment in which cultivation-independent, 16S rRNA gene sequence-based methods were applied to obtain information on the prokaryote community structure, including Archaea as well as Bacteria (Benlloch et al., 1995, 1996; Rodríguez-Valera et al., 1999). These studies identified the phylotype now known to belong to the square archaeon *H. walsbyi*, which is one of the main components of the archaeal community in most saltern crystallizer ponds worldwide (Bolhuis et al., 2004; Burns et al., 2007). The same phylotype was also abundantly found in the Eilat salterns brines (Rodríguez-Valera et al., 1999). These early studies were later expanded when more powerful methods, including denaturing gradient gel electrophoresis, became available (Benlloch et al., 2001, 2002). Different other fingerprinting techniques were applied to the study of the microbiota of the Alicante salterns such as 5S rRNA fingerprints (Casamayor et al., 2000) and ribosomal internal spacer analysis and terminal-restriction fragment length polymorphism (T-RFLP) (Casamajor et al., 2002). Analysis of melting profiles and reassociation kinetics of the DNA extracted from the ponds in combination with T-RFLP showed the contribution of *Haloquadratum*, which has a much lower guanine + cytosine percentage in its DNA than most other halophilic Archaea. In ponds with 22%, 32%, and 37% salinity the community genome was estimated to be 7, 13, and 4 times as complex as the *Escherichia coli* genome (Øvreås et al., 2003). An environmental genomics study of *Haloquadratum* in Alicante showed the existence of a large pool of accessory genes in the community, so that a large pan-genome of *Haloquadratum* is present even in one crystallizer pond (Legault et al., 2006).

The existence of the rod-shaped red extremely halophilic representative of the *Bacteroidetes* now known as *Salinibacter ruber* (Antón et al., 2002) was first indicated during fluorescence in situ hybridization (FISH) analysis of the prokaryotic community inhabiting the Alicante crystallizer ponds (Antón et al., 1999). Further sequencing work and FISH studies showed *Salinibacter* to represent 5%–25% of the total prokaryote community (Antón et al., 2000). Presence of *Salinibacter* in the Santa Pola salterns could also be directly demonstrated by the presence of its carotenoid pigment salinixanthin, contributing ~5% of the total carotenoids present in the pigments extracted from the community (Oren and Rodríguez-Valera, 2001).

Microautoradiography following incubation of Alicante brine with different radiolabeled substrates, combined with FISH suggested that *Haloquadratum* metabolizes amino acids and acetate, while *Salinibacter* did not take up amino acids; glycerol was not used by either organism in these experiments (Rosselló-Mora et al., 2003), a result not confirmed in later pure culture studies with *Salinibacter*.

In recent years, state-of-the-art metagenomics studies performed in the Alicante salterns and the Huelva salterns on the Atlantic coast of Spain showed not only many similarities in the community structure of ponds of the same salinity, but also local differences (Fernández et al., 2014a). Such metagenomics studies, including the previously often neglected intermediate salinity ponds, showed the presence of interesting novel types of prokaryotes: the nanohaloarchaeota abundant at high salinity and leading a photoheterotrophic and polysaccharide-degrading lifestyle, a new group of euryarchaeota, and a group of yet uncultured actinobacteria in a pond of 19% salinity (Ghai et al., 2011; Fernández et al., 2014b). Such metagenomic studies also led to the recognition of *Spiribacter salinus*, a member of a new genus recently isolated from the Huelva saltern (León et al., 2014).

The Santa Pola, Alicante salterns also yielded much information on the diversity of bacteriorhodopsin and other retinal protein genes derived from halophilic Archaea as well as Bacteria present in ponds of different salinities (Papke et al., 2003; Gomariz et al., 2015). Genes for bacteriorhodopsin and related proteins were also detected in an intermediate salinity (13%) pond, suggesting that also in these relatively low-salinity ponds light may be a significant, and thus far overlooked energy source for the prokaryote (photo)heterotrophic community (Fernández et al., 2014b).

An in-depth study of the distribution of planktonic photoautotrophic microorganisms, prokaryotic as well as eukaryotic, along the salinity gradient was made using microscopy, flow cytometry, pigment analysis, and DNA-based methods. A major discontinuity in the distribution patterns was observed between 15% and 22% salinity, where the mixed community of dinoflagellates and diatoms found at the lower salinities was replaced by a community of *Dunaliella* and cyanobacteria adapted to higher salt concentrations (Estrada et al., 2004).

The Alicante salterns also served as a model system to investigate the presence and activity of viruses regulating the community of Archaea and Bacteria. Virus-like particles are abundantly found, typically in numbers around one order of magnitude higher than the prokaryote numbers. Electron microscopic observations showed that 1%–10% of the square flat *Haloquadratum* cells contained visible viruses, and up to ~200 virus-like particles were counted in lysing cells (Guixa-Boixareu et al., 1996). A metatranscriptomic analysis of the halophilic viral communities in a pond of 32% salinity showed the highest expression of putative haloviruses against high G + C Archaea and *Salinibacter*. The effect of UV and dilution stress on the virus community was also tested; the Archaea appear to be more susceptible to viral attacks than the Bacteria when subjected to these stress conditions (Santos et al., 2011).

The intensive studies, cultivation dependent as well as cultivation independent, performed in the Santa Pola, Alicante saltern as a model for studying the microbial ecology of hypersaline environments, were recently summarized in two review papers (Ventosa et al., 2014, 2015).

3 Pigment studies

The colorful layered benthic mats and the pink-red brines of saltern ponds are excellent objects for the study of the occurrence and functioning of microbial pigments in different types of halophiles. The pigments present in the differently colored layers in the endoevaporitic microbial mat within the gypsum crust in the Eilat salterns (Oren et al., 1995, 2009a) (Fig. 1) as well as in other saltern systems (Caumette, 1993; Caumette et al., 1994), include chlorophyll *a* of cyanobacteria, bacteriochlorophyll *a* of anoxygenic phototrophic purple bacteria, and different carotenoid pigments associated with these groups. Emission spectroscopy and kinetic fluorometry studies of the phototrophic communities in the Eilat ponds also yielded indications for the presence of green, chlorosome-containing anoxygenic phototrophs at a salinity of ~20% (Prášil et al., 2009), but the nature of these organisms remains to be elucidated. The upper orange-colored layer of the Eilat gypsum crust that contains dense communities of unicellular *Aphanothece*-like cyanobacteria also has a high content of different mycosporine-like amino acids that may not only act as UV protectants but may also be involved in osmotic protection (Oren, 1997). However, glycine betaine is the major osmotic solute in the microbial communities in this gypsum crust as shown using Raman spectroscopy and nuclear ^{1}H- and ^{13}C-magnetic resonance (Oren et al., 2013).

At least four different pigments may contribute to the pink-orange color of saltern crystallizer brines: *Dunaliella*-derived β-carotene; α-bacterioruberin and derivatives that are the main pigments of the halophilic Archaea; bacteriorhodopsin that can be present as well in halophilic Archaea; and salinixanthin, the carotenoid pigment of *Salinibacter*. Although quantitatively β-carotene is present in the largest amounts (see also Table 1 in Oren et al., 2016), its contribution to the overall coloration of the Eilat saltern brine is minor, due to the fact that the pigment is concentrated in dense granules inside the chloroplast (Oren and Dubinsky, 1994), so that bacterioruberin derivatives contribute most of the color to the brines. Salinixanthin could not be directly demonstrated in cell pellets from the Eilat salterns, but it was a minor component of the carotenoid pigments in the Alicante salt ponds (Oren and Rodríguez-Valera, 2001). Presence of bacteriorhodopsin, the retinal-containing membrane-bound proton pump, could be demonstrated in the crystallizer ponds in Eilat at a concentration of ~3.6 nmol/L. The respiration by the heterotrophic prokaryotic community in the Eilat brine was ~40% lower in the light than in the dark, suggesting that the Archaea that harbor bacteriorhodopsin can derive a major part of their energy from light (Oren et al., 2016).

The Eilat gypsum crust and crystallizer brines were shown to be excellent model systems to test handheld Raman spectrometers for fast screening of microbial pigments (Jehlička and Oren, 2013; Jehlička et al., 2013).

4 Lipids as biomarkers for specific groups of Archaea and Bacteria

The high densities of the microbial communities in saltern crystallizer brines and in the benthic microbial mats in the evaporation ponds at lower salinities enable the easy collection of sufficient material for lipid studies.

Fatty acid analysis of the cyanobacteria that dominate the microbial communities in the evaporitic gypsum crusts in the Eilat saltern yielded important information about the potentially anaerobic life style, at least temporarily, of the layer of *Phormidium*-like filamentous cyanobacteria found below the upper layer of orange-pigmented *Aphanothece*-like unicellular cyanobacteria. The green filamentous cyanobacteria did not contain polyunsaturated fatty acids, and the location of the double bonds in the monounsaturated fatty acids indicated activity of the "bacterial," oxygen-independent pathway for their biosynthesis (Ionescu et al., 2007; Oren et al., 2009a).

The polar lipid patterns, spatial as well as temporal, of the planktonic biota along the salinity gradient of the Eilat salterns were compared to those of the communities in the salterns in San Francisco Bay. Significant differences were found, correlating with the oligotrophic nature of the Eilat salterns and the much higher nutrient concentrations of the Californian salterns (Litchfield and Oren, 2001; Litchfield et al., 2000, 2009).

Analysis of the polar lipids of the Eilat crystallizer ponds, dominated by *Haloquadratum*, enabled the elucidation of the nature of the types of lipids present in this unusual flat square archaeon long before the organism was isolated and cultivated so that its lipids could be studied in axenic culture (Oren, 1994; Oren et al., 1996; Burns et al., 2007).

5 Community metabolism in the saltern brines and sediments

The Eilat salterns have proven to be the ideal environment to obtain information about the in situ activities of halophilic and halotolerant heterotrophic microorganisms along the salinity gradient. The use of specific inhibitors affecting either the archaeal or the bacterial component of the community to obtain information about the relative contribution of either group to the overall heterotrophic activity was pioneered in Eilat: taurocholate and other bile acids that at low concentrations cause lysis of the Archaea without greatly affecting the Bacteria, aphidicolin as a specific inhibitor of DNA replication in the halophilic Archaea, anisomycin that abolishes protein synthesis in halophilic Archaea, and antibiotics such as chloramphenicol that are more or less specific inhibitors of the bacterial protein synthesis machinery. These studies showed that below $\sim$25% salinity most activity can be attributed to the Bacteria, while above that salinity the Archaea take over (Oren, 1990a,b,c).

Much of the information on the activity of *Dunaliella* as the main or sole primary producer in hypersaline environments worldwide is based on studies using saltern ponds as model systems (Oren, 2009). Glycerol is the organic osmotic solute by *Dunaliella salina*, and it can therefore be expected to be a key nutrient for the heterotrophic communities in the ponds of the highest salinities (Elevi Bardavid et al., 2008). Therefore the availability, uptake, and turnover of glycerol were investigated in the Eilat salterns. Uptake rates were high and turnover times of glycerol were in the order of 2.6–7.2 h in the crystallizer ponds (Oren, 1993). Glycerol was also found to stimulate community respiration as measured by oxygen uptake (Warkentin et al., 2009). When the brines were amended with radiolabeled glycerol, even in micromolar concentrations, not all the glycerol was incorporated into cell material or oxidized to carbon dioxide; about 10% was partially converted to D-lactate and acetate. The lactate was subsequently metabolized further but labeled acetate remained present for long times (Oren and Gurevich, 1994). Therefore acetate metabolism in the ponds was studied in further depth. The affinity of the community for acetate is low, and estimated turnover times for acetate were as long as 127–730 h (Oren, 1995).

The diel activities of different groups of microorganisms in the benthic gypsum crusts in salterns of $\sim$20% salinity were probed with oxygen and sulfide microelectrodes, as well as by estimation of sulfate reduction rates using radiolabeled sulfate (Canfield et al., 2004).

The approach of using taurocholate and specific antibiotics (erythromycin) to assess the contribution of Archaea and Bacteria to the community heterotrophic activity, as first used in Eilat (Oren, 1990b,c) was also applied in two Spanish salterns, including the ones near Alicante (Pedrós-Alió et al., 2000). This study concluded that bacterivory by protozoa is insignificant at salinities above 25%, and that viral lysis is of minor importance throughout the salt concentration gradient. Further studies of the heterotrophic prokaryotic abundance and growth rate in the Santa Pola saltern used addition of different organic and inorganic nutrients and removal of zooplankton by filtration to assess the factors limiting the activities in ponds of 4%, 11%, and 22% salinity. An active, substrate-limited community was shown to exist in the low salinity rage; an active grazer-controlled community in the medium salinity ponds; and a possibly dormant, probably substrate-limited community in the highest salinity ponds (Gasol et al., 2004). A study of primary production, nutrient assimilation, and microzooplankton grazing along a salinity gradient in the Alicante saltern showed ammonium to be the dominant nitrogen source. Above 20% salinity rates of carbon dioxide fixation were very low (Joint et al., 2002).

6 Miscellaneous studies in the Eilat salterns

The abundant occurrence of *Haloquadratum* in the Eilat salterns provided an opportunity to test the possible function of the gas vesicles in this intriguing organism. Contrary to what was generally assumed, the gas vesicles do not cause the square cells to float toward the surface of the brine where more oxygen may be present or more light may be available to drive the light-dependent proton pump bacteriorhodopsin. The gas vesicles provide approximately neutral buoyancy, and flotation rates cannot be significant (Oren et al., 2006). The alternative theory that the gas vesicles, which are mainly located at the edges of the cells, can keep the cells oriented parallel to the brine surface and so increase the amount of light available to be used as energy source, remains to be proven (Oren, 2013).

The Eilat saltern crystallizer ponds also served as a testing ground to evaluate the possible importance of halocins, bacteriocin-like proteins excreted by some halophilic Archaea with an antagonistic action on other members of the group. Dissolved high-molecular-weight compounds (>1 or 5 kDa) in the saltern brine were concentrated by ultrafiltration and the possible inhibitory action of the concentrate was tested on a variety of members of the *Halobacteria* class. No indications were obtained for the presence of significant concentrations of inhibitory halocins in the brine. A similar result was obtained with brines from the Santa Pola salterns and from the salterns in San Francisco Bay (Kis-Papo and Oren, 2000).

7 Selected studies of salterns at other geographical locations

The previous sections highlighted research performed in two very intensively investigated saltern ponds in Israel and in Spain that served as models for hypersaline environments in general and for saltern systems worldwide. Many other coastal salterns and also some inland salterns have been the object of microbiological studies. A full overview of all investigations on the biota of these salterns is outside the scope of this short review. I only present here a few highlights:

- Cyanobacterial mats and benthic gypsum crusts similar to those described from the Eilat saltern ponds (Oren et al., 1995, 2009a) were studied in the Salins-de-Giraud salterns on the Mediterranean coast of France. Aspects studied at that site include the zonation of phototrophs, rates of oxygen production and consumption, sulfate reduction, and sulfide oxidation (Caumette, 1993; Caumette et al., 1994), and the structure of the prokaryotic community of the anoxic sediments as investigated using cultivation-independent techniques (Mouné et al., 2002).
- The community of halophilic prokaryotes in anaerobic sediments from a solar saltern on the island of Mallorca was assessed based on 16S rRNA gene sequences and sequences of *dsrAB*, a gene involved in dissimilatory sulfate reduction. Sulfate reduction rates in the sediments were also measured. A novel branch of putative methanogenic Archaea as yet without cultured representatives (candidate division MBSl-1) was identified in the Mallorca saltern sediments (López-López et al., 2010).
- The diversity of the flat square archaeon *H. walsbyi*, which is a major component of the microbial community of saltern crystallizers worldwide, was studied in three, geographically distant, Australian salterns. In spite of the large geographical distance separating the salterns, the result of the 16S rRNA gene clone library study showed a few minor differences between *Haloquadratum* from different locations (Oh et al., 2010).
- The microbial diversity along a salinity gradient (18%, 37%, and 38% salt) in the Guerrero Negro, Baja California, Mexico salterns, including communities of *Haloquadratum* and *Salinibacter*, was assessed based on 16S rRNA and *bop* (bacterio-opsin) genes (Dillon et al., 2013).
- The salterns of Sečovlje on the border between Slovenia and Croatia, a system that has been operated using traditional salt making techniques since the Middle Ages, is unusual as *Haloquadratum* is not prominently present there. Cultivation-independent approaches showed the presence of a high diversity of other Archaea, as well as a great diversity of bacteriorhodopsin genes (Pašić et al., 2005, 2007).
- Application of state-of-the-art MALDI-TOF/MS "lipidomics" yielded detailed information on the lipids, including specific biomarkers for certain groups of halophilic Archaea and Bacteria, in the salterns of Margherita di Savoia on the Adriatic coast of Italy (Lopalco et al., 2011).

8 A laboratory model of a solar saltern

For research on salterns by scientists who do not work in the close vicinity of coastal salt production plants, Giani et al. (1989) developed a laboratory-scale model of a solar saltern. It consisted of a series of Plexiglas ponds, measuring in total

~85 × 250 cm, with one "large" evaporation pond of ~85 × 85 cm and 18 "ponds" of ~27.5 × 27.5 cm with directional flow from pond to pond. The "ponds" were filled with a 5-cm thick layer of sediment collected from a saltern in Bretagne, France, and the water level was kept constant at 2 cm depth above the sediment. Artificial seawater (salinity 3%) was fed into the first evaporation by a peristaltic pump. The system was illuminated by two 2000 W lamps at 1 m distance, and temperature and humidity were controlled at 30°C and 40%, respectively. A stable salinity gradient was established in about 1 week. Opaque or black Plexiglass sheets were placed over some of the "ponds" mats to study the effect of light intensity on the development of microbial mats. After 6 months' operation the regime was changed to 25°C and 50% humidity during 14-h light periods (14 h) and 12°C and 75% humidity during dark periods (10 h). The microbial mats in the "ponds" consisted of diatoms, cyanobacteria, and *Chloroflexus*-like filamentous bacteria of types similar to those in the outdoor salterns. The brine in the most saline "ponds" turned red. No *Dunaliella* cells were observed in the brine, and therefore the red color was presumably caused by halophilic Archaea (Giani et al., 1989).

9 Final comments

The salterns of Eilat and Alicante, highlighted in this review, have been the object of interdisciplinary studies thanks to special research programs that have brought scientists from different countries together to study different aspects of the microbiology of the salterns. The "MIDAS" workshop, funded by the European Community, held at the Santa Pola Salterns in May 1999 and the "GAP" (Group of Aquatic Productivity) workshop that took place in Eilat in March 2008 did much to advance our knowledge on the salterns' biota and their activities. The "Midas" workshop mainly assessed the diversity of the microbial communities in the brines along the salt gradient, as well as their activities. Research performed during the GAP workshop included studies on classic morphology-based taxonomy of phototrophic organisms, emission spectroscopy and kinetic fluorimetry, molecular cultivation-independent studies of sulfate-reducing bacteria and methanogens, and activity measurements using microelectrodes and optodes. Many of the papers cited in this review describe the results of the research performed during these two workshops. The Eilat and the Santa Pola salterns have thus become some of the most intensively studied hypersaline environments worldwide (Oren et al., 2009b; Ventosa et al., 2014).

Editors' Note
Salty Environments and Astrobiology
NASA announced recently that the planet Mars contains body of salty and brine water under its ice layer, which lies beneath the southern ice cap (Orosei et al., 2018). It is known that wherever there is water might be good chances of life.

Similar salty oceans under the ice layer has been proposed for *Europa* (satellite of Jupiter) (Hand and Carlson, 2015) and for *Enceladus* (moon of Saturn) (Sherwood, 2016) and other satellites around them.

The Vostok ocean has been located 4 km under the icy layer of Antarctica (Vasilev et al., 2016). The Russian scientists found remainders of life in this salty lake from microorganisms under an icy ocean sealed for millions of years. They claim that the DNA from the Vostok area does not match any known species in world data (Bulat, 2016).

Places in Earth which contain high level of salt and resembling Mars (Astrobiology).

1. *Dead Sea* in Israel is a home for extremophilic salt lovers (microorganisms such as Archaea, bacteria, and sometime green algae). This salty lake is an analog to Mars oceans in the geological past.
2. *Rio Tinto* in Spain is a salty red river containing acidophilic microorganisms such as acidophilic bacteria and algae.
3. *Yellowstone National Park* [Wyoming, and other States, USA] contains salty hot acidic springs, which are assumed to have sources of early life. Containing bacteria and thermo-acidic algae (as *Cyanidium caldarium*).
4. *Atacama Desert* in Chile resembles salty badlands, heavily eroded dry barren land, hot arid area, and is analog of ancient Mars.
5. *McMurdo dry valley* in Antarctica. One of the driest and coldest deserts of Earth (Analog to Mars). It includes the saline lake Vida, There occur endolithic photosynthesis.
6. West Australia sea is home of most ancient rocks on Earth (Stromatolites of about 3.8 BYA).

Conclusion

The salty environment has strong connection to Astrobiology such as the Origin of Life, evolution, and halophiles (Gordon et al., 2017). It is hoped that also some planets (such as Mars) and Satellites (such as Europa and Enceladus and others) will contain in their subicy salty liquid water some living microorganisms. Since Origin of Life seems to have taken place in salt liquid environments (perhaps sub ocean vents, or hot springs), therefore all organisms need and use salt in their life.

References

Bulat, S.A., 2016. Microbiology of the subglacial Lake Vostok: first results of borehole-frozen lake water analysis and prospects for searching for lake inhabitants. Philos. Trans. R. Soc. A-Math. Phys. Eng. Sci. 374(2059), #20140292.

Gordon, R., Hanczyc, M.M., Denkov, N.D., Tiffany, M.A., Smoukov, S.K., 2017. Chapter 18: Emergence of polygonal shapes in oil droplets and living cells: the potential role of tensegrity in the origin of life, in: Gordon, R., Sharov, A.A. (Eds.), Habitability of the Universe Before Earth [Volume 1 in series: Astrobiology: Exploring Life on Earth and Beyond, eds. Pabulo Henrique Rampelotto, Joseph Seckbach & Richard Gordon]. Elsevier B.V., Amsterdam, pp. 427–490.

Hand, K.P., Carlson, R.W., 2015. Europa's surface color suggests an ocean rich with sodium chloride. Geophys. Res. Lett. 42(9), 3174–3178.

Orosei, R., Lauro, S.E., Pettinelli, E., Cicchetti, A., Coradini, M., Cosciotti, B., Di Paolo, F., Flamini, E., Mattei, E., Pajola, M., Soldovieri, F., Cartacci, M., Cassenti, F., Frigeri, A., Giuppi, S., Martufi, R., Masdea, A., Mitri, G., Nenna, C., Noschese, R., Restano, M., Seu, R., 2018. Radar evidence of subglacial liquid water on Mars.

Sherwood, B., 2016. Strategic map for exploring the ocean-world Enceladus. Acta Astron. 126, 52–58.

Vasilev, N.I., Dmitriev, A.N., Lipenkov, V.Y., 2016. Results of the 5G borehole drilling at Russian Antarctic station "Vostok" and researches of ice cores [Russian]. J. Mining Inst. 218, 161–171.

References

Antón, J., Llobet-Brossa, E., Rodríguez-Valera, F., Amann, R., 1999. Fluorescence *in situ* hybridization analysis of the prokaryotic community inhabiting crystallizer ponds. Environ. Microbiolol. 1, 517–523.

Antón, J., Rosselló-Mora, R., Rodríguez-Valera, F., Amann, R., 2000. Extremely halophilic *Bacteria* in crystallizer ponds from solar salterns. Appl. Environ. Microbiol. 66, 3052–3057.

Antón, J., Oren, A., Benlloch, S., Rodríguez-Valera, F., Amann, R., Rosselló-Mora, R., 2002. *Salinibacter ruber* gen. nov., sp. nov., a novel extreme halophilic member of the *Bacteria* from saltern crystallizer ponds. Int. J. Syst. Evol. Microbiol. 52, 485–491.

Benlloch, S., Martínez-Murcia, A.J., Rodríguez-Valera, F., 1995. Sequencing of bacterial and archaeal 16S rRNA genes directly amplified from a hypersaline environment. Syst. Appl. Microbiol. 18, 574–581.

Benlloch, S., Acinas, S.G., Martínez-Murcia, A.J., Rodríguez-Valera, F., 1996. Description of prokaryotic biodiversity along the salinity gradient of a multipond saltern by direct PCR amplification of 16S rDNA. Hydrobiologia 329, 19–31.

Benlloch, S., Acinas, S.G., Antón, J., López-López, A., Luz, S.P., Rodríguez-Valera, F., 2001. Archaeal biodiversity in crystallizer ponds from a solar saltern: culture versus PCR. Microb. Ecol. 41, 12–19.

Benlloch, S., López-López, A., Casamayor, E.O., Øvreås, L., Goddard, V., Daae, F.L., Smerdon, G., Massana, R., Joint, I., Thingstad, F., Pedrós-Alió, C., Rodríguez-Valera, F., 2002. Prokaryotic genetic diversity throughout the salinity gradient of a coastal solar saltern. Environ. Microbiol. 4, 349–360.

Bolhuis, H., te Poele, E.M., Rodríguez-Valera, F., 2004. Isolation and cultivation of Walsby's square archaeon. Environ. Microbiol. 6, 1287–1291.

Burns, D.G., Janssen, P.H., Itoh, T., Kamekura, M., Li, Z., Jensen, G., Rodríguez-Valera, F., Bolhuis, H., Dyall-Smith, M.L., 2007. *Haloquadratum walsbyi* gen. nov., sp. nov., the square haloarchaeon of Walsby, isolated from saltern crystallizers in Australia and Spain. Int. J. Syst. Evol. Microbiol. 57, 387–392.

Butinar, L., Santos, S., Spencer-Martins, I., Oren, A., Gunde-Cimerman, N., 2005. Yeast diversity in hypersaline habitats. FEMS Microbiol. Lett. 244, 229–234.

Canfield, D.E., Sørensen, K.B., Oren, A., 2004. Biogeochemistry of a gypsum-encrusted microbial ecosystem. Geobiology 2, 133–150.

Casamajor, E.O., Massana, R., Benlloch, S., Øvrås, L., Díez, B., Goddard, V.J., Gasol, J.M., Joint, I., Rodríguez-Valera, F., Pedrós-Alió, C., 2002. Changes in archaeal, bacterial and eukaryal assemblages along a salinity gradient by comparison of genetic fingerprinting methods in a multipond solar saltern. Environ. Microbiol. 4, 338–348.

Casamayor, E.O., Calderón-Paz, J.I., Pedrós-Alió, C., 2000. 5S rRNA fingerprints of marine bacteria, halophilic archaea and natural prokaryotic assemblages along a salinity gradient. FEMS Microbiol. Ecol. 34, 113–119.

Caumette, P., 1993. Ecology and physiology of phototrophic bacteria and sulfate-reducing bacteria in marine salterns. Experientia 49, 473–481.

Caumette, P., Matheron, R., Raymond, N., Relexans, J.-C., 1994. Microbial mats in the hypersaline ponds of Mediterranean salterns (Salins-de-Giraud, France). FEMS Microbiol. Ecol. 13, 273–286.

Dillon, J.G., Carlin, M., Gutierrez, A., Nguyen, V., McLain, N., 2013. Patterns of microbial diversity along a salinity gradient in the Guerrero Negro solar saltern, Baja CA sur, Mexico. Front. Microbiol. 4, 399.

Elevi Bardavid, R., Khristo, P., Oren, A., 2008. Interrelationships between *Dunaliella* and halophilic prokaryotes in saltern crystallizer ponds. Extremophiles 12, 5–14.

Estrada, M., Henriksen, P., Gasol, J.M., Casamayor, E.O., Pedrós-Alió, C., 2004. Diversity of planktonic photoautotrophic microorganisms along a salinity gradient as depicted by microscopy, flow cytometry, pigment analysis and DNA-based methods. FEMS Microbiol. Ecol. 49, 281–293.

Fernández, A.B., Vera-Gargallo, B., Sánchez-Porro, C., Ghai, R., Papke, R.T., Rodriguez-Valera, F., Ventosa, A., 2014a. Comparison of prokaryotic community structure from Mediterranean and Atlantic saltern concentrator ponds by a metagenomic approach. Front. Microbiol. 5, 196.

Fernández, A.B., Ghai, R., Martin-Cuadrado, A.-B., Sánchez-Porro, C., Rodriguez-Valera, F., Ventosa, A., 2014b. Prokaryotic taxonomic and metabolic diversity of an intermediate salinity hypersaline habitat assessed by metagenomics. FEMS Microbiol. Ecol. 88, 623–635.

Gasol, J.M., Casamayor, E.O., Joint, I., Garde, K., Gustavson, K., Benlloch, S., Díez, B., Schauer, M., Massana, R., Pedrós-Alió, C., 2004. Control of heterotrophic prokaryotic abundance and growth rate in hypersaline planktonic environments. Aquat. Microb. Ecol. 34, 193–206.

Ghai, R., Pašić, L., Fernández, A.B., Martin-Cuadrado, A.-B., Megumi Mizuno, C., McMahon, K.D., Papke, R.T., Stepanauskas, R., Rodriguez-Brito, B., Rohwer, F., Sánchez-Porro, C., Ventosa, A., Rodríguez-Valera, F., 2011. New abundant microbial groups in aquatic hypersaline environments. Sci. Rep. 1, 135.

Giani, D., Seeler, J., Giani, L., Krumbein, W.E., 1989. Microbial mats and physicochemistry in a saltern in the Bretagne (France) and in a laboratory scale model. FEMS Microbiol. Ecol. 62, 151–162.

Gomariz, M., Martínez-García, M., Santos, F., Constantino, M., Meseguer, I., Antón, J., 2015. Retinal-binding proteins mirror prokaryotic dynamics in multipond solar salterns. Environ. Microbiol. 17, 514–526.

Guixa-Boixareu, N., Calderón-Paz, J.I., Heldal, M., Bratbak, G., Pedrós-Alió, C., 1996. Viral lysis and bacterivory as prokaryotic loss factors along a salinity gradient. Aquat. Microb. Ecol. 11, 215–227.

Ionescu, D., Lipski, A., Altendorf, K., Oren, A., 2007. Characterization of the endoevaporitic microbial communities in a hypersaline gypsum crust by fatty acid analysis. Hydrobiologia 576, 15–26.

Javor, B.J., 1983. Planktonic standing crop and nutrients in a saltern ecosystem. Limnol. Oceanogr. 28, 153–159.

Javor, B., 1989. Hypersaline Environments. Microbiology and Biogeochemistry. Springer-Verlag, Berlin.

Javor, B.J., 2002. Industrial microbiology of solar salt production. J. Ind. Microbiol. Biotechnol. 28, 42–47.

Jehlička, J., Oren, A., 2013. Use of a handheld Raman spectrometer for fast screening of microbial pigments in cultures of halophilic microorganisms and in microbial communities in hypersaline environments in nature. J. Raman Spectrosc. 44, 1285–1291.

Jehlička, J., Edwards, H.G.M., Oren, A., 2013. Bacterioruberin and salinixanthin carotenoids of extremely halophilic Archaea and Bacteria: a Raman spectroscopic study. Spectrochim. Acta A 106, 99–103.

Joint, I., Henriksen, P., Garde, K., Riemann, B., 2002. Primary production, nutrient assimilation and microzooplankton grazing along a hypersaline gradient. FEMS Microbiol. Ecol. 39, 245–257.

Kis-Papo, T., Oren, A., 2000. Halocins: are they involved in the competition between halobacteria in saltern ponds? Extremophiles 4, 35–41.

Legault, B.A., Lopez-Lopez, A., Alba-Casado, J.C., Doolittle, W.F., Bolhuis, H., Rodriguez-Valera, F., Papke, R.T., 2006. Environmental genomics of "Haloquadratum walsbyi" in a saltern crystallizer indicates a large pool of accessory genes in an otherwise coherent species. BMC Genomics 7, 171.

León, M.J., Fernández, A.B., Ghai, R., Sánchez-Porro, C., Rodriguez-Valera, F., Ventosa, A., 2014. From metagenomics to pure culture: isolation and characterization of the moderately halophilic bacterium Spiribacter salinus gen. nov., sp. nov. Appl. Environ. Microbiol. 80, 3850–3857.

Litchfield, C.D., Oren, A., 2001. Polar lipids and pigments as biomarkers for the study of the microbial community structure of solar salterns. Hydrobiologia 466, 81–89.

Litchfield, C.D., Irby, A., Kis-Papo, T., Oren, A., 2000. Comparisons of the polar lipid and pigment profiles of two solar salterns located in Newark, California, U.S.A., and Eilat, Israel. Extremophiles 4, 259–265.

Litchfield, C.D., Oren, A., Irby, A., Sikaroodi, M., Gillevet, P.M., 2009. Temporal and salinity impacts on the microbial diversity at the Eilat, Israel solar salt plant. Global NEST J. 11, 86–90.

Lopalco, P., Lobasso, S., Baronio, M., Angelini, R., Corcelli, A., 2011. Impact of lipidomics on the microbial world of hypersaline environments. In: Ventosa, A., Oren, A., Ma, Y. (Eds.), Halophiles and Hypersaline Environments. Springer-Verlag, Berlin, pp. 123–135.

López-López, A., Yarza, P., Richter, M., Suárez-Suárez, A., Antón, J., Niemann, H., Rosselló-Móra, R., 2010. Extremely halophilic microbial communities in anaerobic sediments from a solar saltern. Environ. Microbiol. Rep. 2, 258–271.

Mouné, S., Caumette, P., Matheron, R., Willison, J.C., 2002. Molecular sequence analysis of prokaryotic diversity in the anoxic sediments underlying cyanobacterial mats of two hypersaline ponds in Mediterranean salterns. FEMS Microbiol. Ecol. 44, 117–130.

Oh, D., Porter, K., Russ, B., Burns, D., Dyall-Smith, M., 2010. Diversity of Haloquadratum and other haloarchaea in three, geographically distant, Australian saltern crystallizer ponds. Extremophiles 14, 161–169.

Oren, A., 1990a. Thymidine incorporation in saltern ponds of different salinities: estimation of in situ growth rates of halophilic archaeobacteria and eubacteria. Microb. Ecol. 19, 43–51.

Oren, A., 1990b. Estimation of the contribution of halobacteria to the bacterial biomass and activity in solar salterns by the use of bile salts. FEMS Microbiol. Ecol. 73, 41–48.

Oren, A., 1990c. The use of protein synthesis inhibitors in the estimation of the contribution of halophilic archaebacterial to bacterial activity in hypersaline environments. FEMS Microbiol. Ecol. 73, 187–192.

Oren, A., 1993. Availability, uptake, and turnover of glycerol in hypersaline environments. FEMS Microbiol. Ecol. 12, 15–23.

Oren, A., 1994. Characterization of the halophilic archaeal community in saltern crystallizer ponds by means of polar lipid analysis. Int. J. Salt Lake Res. 3, 15–29.

Oren, A., 1995. Uptake and turnover of acetate in hypersaline environments. FEMS Microbiol. Ecol. 18, 75–84.

Oren, A., 1997. Mycosporine-like amino acids as osmotic solutes in a community of halophilic cyanobacteria. Geomicrobiol J. 14, 233–242.

Oren, A., 2002. Halophilic Microorganisms and their Environments. Kluwer Scientific Publishers, Dordrecht.

Oren, A., 2005. Saltern crystallizer ponds as field laboratories for the study of extremely halophilic Archaea and Bacteria. In: Proceedings of the International Symposium on Extremophiles and their Applications, Tokyo, pp. 282–289. http:/www.jstage.jst.go.jp/article/isea/2005/0/282/_pdf.

Oren, A., 2009. Saltern evaporation ponds as model systems for the study of primary production processes under hypersaline conditions. Aquat. Microb. Ecol. 56, 193–204.

Oren, A., 2010. Thoughts on the "missing link" between saltworks biology and solar salt quality. Global NEST J. 12, 417–425.

Oren, A., 2013. The function of gas vesicles in halophilic Archaea and Bacteria: theories and experimental evidence. Life 3 (1), 1–20. https://doi.org/10.3390/life3010001.

Oren, A., Dubinsky, Z., 1994. On the red coloration of saltern crystallizer ponds. II. Additional evidence for the contribution of halobacterial pigments. Int. J. Salt Lake Res. 3, 9–13.

Oren, A., Gurevich, P., 1994. Production of D-lactate, acetate, and pyruvate from glycerol in communities of halophilic archaea in the Dead Sea and in saltern crystallizer ponds. FEMS Microbiol. Ecol. 14, 147–156.

Oren, A., Rodríguez-Valera, F., 2001. The contribution of halophilic Bacteria to the red coloration of saltern crystallizer ponds. FEMS Microbiol. Ecol. 36, 123–130.

Oren, A., Kühl, M., Karsten, U., 1995. An endoevaporitic microbial mat within a gypsum crust: zonation of phototrophs, photopigments, and light penetration. Mar. Ecol. Prog. Ser. 128, 151–159.

Oren, A., Duker, S., Ritter, S., 1996. The polar lipid composition of Walsby's square bacterium. FEMS Microbiol. Lett. 138, 135–140.

Oren, A., Pri-El, N., Shapiro, O., Siboni, N., 2006. Buoyancy studies in natural communities of square gas-vacuolate archaea in saltern crystallizer ponds. Saline Syst. 2, 4.

Oren, A., Sørensen, K.B., Canfield, D.E., Teske, A.P., Ionescu, D., Lipski, A., Altendorf, K., 2009a. Microbial communities and processes within a hypersaline gypsum crust in a saltern evaporation pond (Eilat, Israel). Hydrobiologia 626, 15–26.

Oren, A., Bina, D., Ionescu, D., Prášil, O., Řeháková, K., Schumann, R., Sørensen, K., Warkentin, M., Woelfel, J., Zapomělová, E., 2009b. Saltern evaporation ponds as model systems for the study of microbial processes under hypersaline conditions – an interdisciplinary study of the salterns of Eilat, Israel. In: Lekkas, T.D., Korovessis, N.A. (Eds.), Proceedings of the 2nd Conference on the Ecological Importance of Solar Saltworks, Merida, Mexico, pp. 20–29.

Oren, A., Elevi Bardavid, R., Kandel, N., Aizenshtat, Z., Jehlička, J., 2013. Glycine betaine is the main organic osmotic solute in a stratified microbial community in a hypersaline evaporitic gypsum crust. Extremophiles 17, 445–451.

Oren, A., Abu-Ghosh, S., Argov, T., Kara-Ivanov, E., Shitrit, D., Volpert, A., Horwitz, R., 2016. Expression and functioning of retinal-based proton pumps in a saltern crystallizer brine. Extremophiles 20, 69–77.

Øvreås, L., Daae, F.L., Torsvik, V., Rodríguez-Valera, F., 2003. Characterization of microbial diversity in hypersaline environments by melting profiles and reassociation kinetics in combination with terminal restriction fragment length polymorphism (T-RFLP). Microb. Ecol. 46, 291–301.

Papke, R.T., Douady, C.J., Doolittle, W.F., Rodríguez-Valera, F., 2003. Diversity of bacterio-rhodopsins in different hypersaline waters from a single Spanish saltern. Environ. Micro-biol. 5, 1039–1045.

Pašić, L., Galán Bartual, S., Poklar Ulrih, N., Grabnar, M., Herzog Velikonja, B., 2005. Diver-sity of halophilic archaea in the crystallizers of an Adriatic solar saltern. FEMS Microbiol. Ecol. 54, 491–498.

Pašić, L., Poklar Ulrih, N., Črnigoj, M., Grabnar, M., Herzog Velikonja, B., 2007. Haloarchaeal communities in the crystallizers of two adriatic solar salterns. Can. J. Microbiol. 53, 8–18.

Pedrós-Alió, C., Calderón-Paz, J.I., MacLean, M.H., Medina, G., Marassé, C., Gasol, J.M., Guixa-Boixereu, N., 2000. The microbial food web along salinity gradients. FEMS Micro-biol. Ecol. 32, 143–155.

Prášil, O., Bína, D., Medová, H., Řeháková, K., Zapomělová, E., Veselá, J., Oren, A., 2009. Emission spectroscopy and kinetic fluorometry studies of phototrophic microbial commu-nities along a salinity gradient in solar saltern evaporation ponds of Eilat, Israel. Aquat. Microb. Ecol. 56, 285–296.

Řeháková, K., Zapomělová, E., Prášil, O., Veselá, J., Medová, H., Oren, A., 2009. Composi-tion changes of phototrophic microbial communities along the salinity gradient in the solar saltern evaporation ponds of Eilat, Israel. Hydrobiologia 636, 77–88.

Rodriguez-Valera, F., Ventosa, A., Juez, G., Imhoff, J.F., 1985. Variation of environmental features and microbial populations with salt concentrations in a multi-pond saltern. Microb. Ecol. 11, 107–115.

Rodríguez-Valera, F., Acinas, S.G., Antón, J., 1999. Contribution of molecular techniques to the study of microbial diversity in hypersaline environments. In: Oren, A. (Ed.), Microbi-ology and Biogeochemistry of Hypersaline Environments. CRC Press, Boca Raton, pp. 27–38.

Rosselló-Mora, R., Lee, N., Antón, J., Wagner, M., 2003. Substrate uptake in extremely hal-ophilic microbial communities revealed by microautoradiography and fluorescence in situ hybridization. Extremophiles 7, 409–413.

Santos, F., Moreno-Paz, M., Meseguer, I., López, C., Rosselló-Mora, R., Parro, V., Antón, J., 2011. Metatranscriptomic analysis of extremely halophilic viral communities. ISME J. 5, 1621–1633.

Sørensen, K.B., Canfield, D.E., Teske, A.P., Oren, A., 2005. Community composition of a hypersaline endoevaporitic microbial mat. Appl. Environ. Microbiol. 71, 7352–7365.

Tkavc, R., Ausec, L., Oren, A., Gunde-Cimerman, N., 2011. Bacteria associated with *Artemia* spp. along the salinity gradient of the solar salterns at Eilat (Israel). FEMS Microbiol. Ecol. 77, 310–321.

Ventosa, A., Fernández, A.B., León, M.J., Sánchez-Porro, C., Rodriguez-Valera, F., 2014. The Santa Pola saltern as a model for studying the microbiota of hypersaline environments. Extremophiles 18, 811–824.

Ventosa, A., de la Haba, R.R., Sánchez-Porro, C., Papke, R.T., 2015. Microbial diversity of hypersaline environments: a metagenomic approach. Curr. Opin. Microbiol. 25, 80–87.

Viver, T., Cifuentes, A., Díaz, S., Rodríguez-Valdecantos, G., González, B., Antón, J., Rosselló-Móra, R., 2015. Diversity of extremely halophilic cultivable prokaryotes in Med-iterranean, Atlantic and Pacific solar salterns: evidence that unexplored sites constitute sources of cultivable novelty. Syst. Appl. Microbiol. 38, 266–275.

Warkentin, M., Schumann, R., Oren, A., 2009. Community respiration studies in saltern crys-tallizer ponds. Aquat. Microb. Ecol. 56, 255–261.

The extremophiles of Great Salt Lake: Complex microbiology in a dynamic hypersaline ecosystem

Bonnie K. Baxter*, Polona Zalar[†]

Department of Biology, Great Salt Lake Institute, Westminster College, Salt Lake City, UT, United States* Department of Biology, University of Ljubljana, Ljubljana, Slovenia[†]

Chapter outline

Model Ecosystems in Extreme Environments. https://doi.org/10.1016/B978-0-12-812742-1.00004-0

1 A natural history of Great Salt Lake

1.1 The Bonneville Basin and formation of Great Salt Lake

The Great Basin is the largest contiguous inland watershed in North America, surrounded by mountain ranges and the Wasatch fault zone (Cohenour and Thompson, 1966). In this setting, Great Salt Lake (GSL) lies in one of the lowest depressions, the Bonneville basin. This terminal lake locale has been the home of four deep lakes over the last 780 thousand years, including the Pleistocene Lake Bonneville, 30 to 13 thousand years ago, covering about 32,000 km^2 of western Utah and extending into eastern Nevada and southern Idaho (Oviatt et al., 1999; Shroder et al., 2016). However, the Bonneville basin primarily held shallow lakes such as GSL, or mudflats and playa, likely over the last several million years (Atwood et al., 2016).

The transition of Lake Bonneville to GSL occurred over just a few thousand years (Atwood et al., 2016). As the last ice age thawed and the Earth warmed, an alluvial fan dam in southern Idaho burst, the water rushing out, leaving scars on the landscape that are still visible today (Shroder et al., 2016). The water evaporated and leaked out rapidly, dropping about 200 m in 2000 years, resulting in the current GSL margins by about 13,000 years ago (Fig. 1). Howard Stansbury led the earliest mapping expedition that captured the expanse of the lake in the nineteenth century (Stansbury, 1855).

1.2 Modern Great Salt Lake

GSL is the largest lake in the western United States, the fourth largest meromictic lake in the world, and the second saltiest lake on Earth next to the Dead Sea (Keck and Hassibe, 1979). The major source of freshwater inflow to GSL comes from three major rivers: the Bear, Weber, and Jordan Rivers, which are part of an extensive drainage (Jones et al., 2009) (Fig. 1). Though the precise margins of GSL vary with seasonal precipitation and drought cycles, it measures approximately 122 km in length and 50 km in width with an average depth of 4.3 m and a maximum depth of 9 m (Keck and Hassibe, 1979; Stephens, 1990). This lake experiences significant seasonal temperature variation, from 0.5°C in January to 26.7°C in July (Crosman and Horel, 2009) and up to 45°C in the shallow margins (Post, 1977) due to its elevation and desert setting.

A critical stop on the Pacific flyway, GSL hosts around 10 million waterbirds (more than 250 species) that spend at least part of the year here, making the lake the most important shorebird site in North America (Bellrose, 1980; Oring et al., 2000; Paul and Manning, 2016; Aldrich and Paul, 2002; Neill et al., 2016). The avian population is sustained by a large biomass of two invertebrate species: brine shrimp (*Artemia franciscana*) and brine flies and their larvae (*Ephydra* spp.) (Verrill, 1869; Packard Jr., 1871; Aldrich, 1912; Collins, 1980; Wurtsbaugh and Gliwicz, 2001; Roberts, 2013). The shrimp spend their entire lifecycle in the briny waters, but only the fly's larvae and pupae are in the brine, while the adult flies fly above the surface

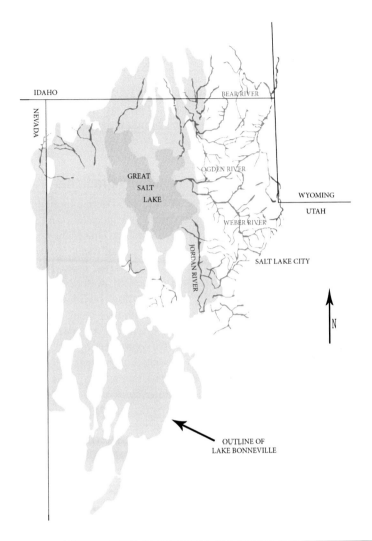

FIG. 1

A map of Great Salt Lake (GSL), Utah, with an overlay of the Pleistocene Lake Bonneville margins for comparison (based on Atwood et al., 2016). Also noted are the positions of Salt Lake City, rivers that feed into Great Salt Lake, and state lines.

Image credit: Kendall Tate-Wright.

and on the shore. The phytoplankton and periphyton of GSL feed these invertebrates (e.g., Felix and Rushforth, 1979; Collins, 1980; Wurtsbaugh and Gliwicz, 2001; Barnes and Wurtsbaugh, 2015). The food web is simple from the macro viewpoint, and the more complex biochemistries of autotrophy and nutrient turnover occur in the microbial communities.

2 Salinity gradients

How salty is GSL? This depends on your vantage point. The salinity ranges from 30 g/L of total dissolved salts to 340 g/L across the various regions of the lake (e.g., State of Utah, 2018a, 2018b; Baxter et al., 2005; Jones et al., 2009; Naftz et al., 2011). Salinity gradients in the lake have been shown to markedly influence the structure and composition of the microbial communities in the water column and benthic regions (Boyd et al., 2014; Meuser et al., 2013).

2.1 Terminal Basin fluctuations and salinity gradients

GSL is a terminal lake. With no outlet, such a closed system experiences cycles of drought and flooding, dependent on temperature, evaporation, and precipitation cycles. Abundant precipitation in contributing watersheds caused the lake to rise four meters from 1983 to 1987 (Stephens, 1990), and years of recent drought set a new historic low in 2016 (USGS, 2018a), the lowest elevation since 1963 (Stephens, 1990). As the lake rises and falls, the salinity of the GSL water column is impacted as high precipitation can dilute the brine, and desiccating conditions can concentrate it. Therefore salinity in the lake varies inversely with lake level. Current upstream water demands coupled with climate change are having an enormous impact on the lake elevation (Fritz, 2014; Deamer, 2016; Wurtsbaugh et al., 2017; USGS, 2018a). Empirical drought reconstruction predicts a catastrophic mega-drought in the southwestern United States, which includes Utah, and suggests that GSL may not recover water elevation gain in the near future (Cook et al., 2015). In fact, saline lakes around the world are evaporating in similar decline (Wurtsbaugh et al., 2017).

Local conditions are impacted by haloclines, where brines of different concentrations do not mix, resulting in brine stratification at a single site (Naftz et al., 2008; Meuser et al., 2013). Unpredictably, weather systems can cause mixing and turnover to alter the salinity in local regions, and mixing of brines can impact the specific ion concentrations of an area (Spencer et al., 1985). GSL studies on brine chemistry and microbiology should be taken in the context of the temporal and spatial conditions that impact the salinity gradient.

2.2 Anthropomorphic impacts

The construction of a rock-filled railroad causeway from 1955 to 1959 bisected GSL and isolated the north arm of the lake, restricting exchange and creating an artificial salinity gradient (Fig. 2) (Madison, 1970; Cannon and Cannon, 2002; Baxter et al., 2005). Within 7 years, this difference was noted as the north arm was approaching saturation, while the south arm, which receives the freshwater input from the watershed, was less saline (Greer, 1971). In recent years, the open waters of south arm of GSL ranged from 110 to 150 g/L (USGS, 2018a) while the north arm is at saturation (280–340 g/L, dependent on the temperature) (e.g., Baxter et al., 2005; Almeida-Dalmet et al., 2015).

FIG. 2

The railroad causeway that separates the north (~30% salt) and south (~15% salt) arms of Great Salt Lake. (A) As seen from space, the causeway bisects the lake. (B) The pink color of the north arm (bottom of photograph) shows an enrichment of carotenoid-containing extremely halophilic archaea.

(A) Image credit: International Space Station, NASA. (B) Image credit: Utah Division of Wildlife, Great Salt Lake Ecosystem Program.

The causeway separating the north and south arms has also caused brine stratification. The formation of a deep brine layer is occasionally observed, due to the denser north arm water seeping through the porous rocks of the causeway and falling below the less saline south arm water (Naftz et al., 2008; Meuser et al., 2013). Recently (December 2016) the breach in the railroad causeway (Fig. 2B) was restored, allowing south arm water to pour into the north arm through a repaired opening large enough to let boats pass (USGS, 2018b). As the lake equilibrates from this action, we may see the formation of new salinity gradients, deep brine layers, and niches for a variety of microbial communities.

Other damming events created critical bird habitats and are part of the Western Hemispheric Shorebird Reserve Network, but they also diverted water from GSL and created a series of interesting microbial environments at a single site. As early as the late 1800s, damming and diversion of water upstream of GSL to create agricultural lands, resulted in other microniches around the lake such as freshwater marshes or brackish pools with a lower salinity. In 1928 the U.S. Congress passed an act to make the Bear River delta a National Wildlife Refuge (US Fish and Wildlife, 2018). Later, federal agencies diked and dammed the Bear River to produce bird habitat there (United States Bureau of Reclamation, 1962). The structures are on the margin of the lake and have been maintained over time, preventing inflow of this water to GSL. For similar reasons, the Farmington Bay Wildlife Management Area was constructed beginning in 1935, which created a brackish bay to the east of Antelope Island in GSL (State of Utah, 2018a). Carp can survive in the 2%–3% salinity, and they are food for large avian species such as Bald Eagles and American White Pelicans.

The GSL mineral extraction industry that produces sodium chloride (road and softener salt), magnesium chloride (for steel production), and potassium sulfate (fertilizer), diverts and dams water (Behrens, 1980; Bingham, 1980; State of Utah, 2018b). The evaporation ponds of these industries, in the north arm and along the edges of the lake, create microbial habits along a salinity gradient as the minerals are concentrated.

3 Unique geochemistry

3.1 Ion concentrations

GSL is a thalassohaline lake, which is an inland saline body of water with ion proportions similar to (but not the same as) the dissolved salts in seawater, indicating it formed from the evaporation of seawater in its history (Oren, 1993). The modern evaporites and solid salts experience cycles of dissolution, bringing a continuous flow of minerals into the brine (Jones et al., 2009). GSL is rich in sodium chloride (Fig. 3A) with an exceptionally high sulfate concentration, which distinguishes GSL from some salt lakes, such as the divalent-rich Dead Sea (Post, 1977; Sturm, 1980; Spencer et al., 1985; Domagalski et al., 1989; Baxter et al., 2005; Jones et al., 2009). The relative ratios of specific ions, for example, 1:1.7 Na^+ to Cl^-, remain somewhat consistent (Table 1) (Gwynn, 1998; Rupke and McDonald, 2012).

3.2 Significant minerals

Beyond the presence of sodium chloride or halite (Fig. 3A), calcium carbonate ($CaCO_3$) precipitation is evident in the GSL lakebed and shores, which are covered with Oolitic sand (Fig. 3B) (Sandberg, 1975; Halley, 1977). Each granule is egg shaped and formed from concentric layers deposited over time around an organic nucleus (such as brine shrimp fecal pellets) (Eardley, 1938; Spencer et al., 1985) (Fig. 3b). Eardley reported the composition to be approximately 84% $CaCO_3$, 5.5% $2MgCO_3 \cdot CaCO_3$, and 5.6% very fine clay. $CaCO_3$ sediments are also evident in the microbialite structures, discussed in Section IV.F. (Chidsey Jr. et al., 2015; Lindsay et al., 2017). Gypsum ($CaSO_4 \cdot 2H_2O$) crystals are abundant in the north arm sediment (Fig. 3C). Measuring 4–13 cm, these crystals are found in the halite-rich clay (Eardley and Stringham, 1952).

3.3 Heavy metal contamination

Utah has a history of mining. Copper, gold, and coal extraction are processes that result in heavy metal by-products for the environment. A large lake in the vicinity of these activities means volatile metals such as mercury can be brought to GSL

FIG. 3

The unusual geochemistry of Great Salt Lake. (A) Halite (sodium chloride) from the north arm, scale bar is 1 cm. (B) Ooids ($CaCO_3$) from the south arm, scale bar is 1 mm. (C) A gypsum crystal ($CaSO_4 \cdot 2H_2O$) from the north arm, scale bar is 1 cm. (D) A petroleum seep at Rozel Bay at the north arm, humans for scale.

Image credits: Great Salt Lake Institute at Westminster College.

Table 1 Ion concentrations of the South arm and the North arm of Great Salt Lake

GSL region	Na^+	K^+	Ca^{2+}	Mg^{2+}	SO_4^{2-}	Cl^-
South arm	59	3.4	0.26	4.6	13	95
North arm	101	6.9	0.28	8.5	22	175

All values in g/L (Utah Geologic Survey, unpublished; Baxter et al., 2005).

through atmospheric deposition (Naftz et al., 2009; Peterson and Gustin, 2008). Deposition of metals may be facilitated by the high concentration of atmospheric reactive chlorine above the lake (Stutz et al., 2002), but others argue against this model (Peterson and Gustin, 2008). Selenium enters through the release of mining discharge waters (Beisnera et al., 2009; Diaz et al., 2009; Naftz et al., 2009). Cobalt, copper, lead, and zinc have also been found in GSL sediments and were attributed to anthropogenic sources (Domagalski et al., 1990). The nature of a terminal lake such as GSL is that (nonvolatile) contaminants do not leave, and geochemical precipitation may store such contaminants in the sediment until turnover occurs (Tayler et al., 1980). The result is that GSL boasts some of the highest concentrations of total mercury measured in an aquatic system in the United States (Naftz et al., 2008). There is demonstrated transfer of mercury through methylated forms (Saxton et al., 2013; Boyd et al., 2017), as well as selenium (Wurtsbaugh, 2007), to the biota of the ecosystem.

3.4 Petroleum seeps

Petroleum is a complex mixture of hydrocarbons and organic compounds, providing a number of possible substrates for microbial metabolism. Rozel Point, in the north arm of GSL, was the subject of oil exploration, in the 1980s until present day, due to visible seeps of high molecular weight petroleum oozing from the surface (Fig. 3D) (Sinninghe-Damsté et al., 1987; Damsté et al., 1987). Ward and Brock (1978) first reported microbial degradation of petroleum, using GSL saltern brine and monitoring the rate of metabolism of ^{14}C-labeled hexadecane. Hydrocarbon turnover was found to decrease sharply with increasing salinity. These investigators failed to cultivate any bacteria nor archaea from the natural oil seeps at Rozel Point (North Arm) although they noted that the phytanes were removed in enrichment cultures implying biodegradation.

4 Microbial diversity

GSL is highly productive despite the reduced solubility of oxygen of hypersaline waters, but we know little about the carbon cycling (Javor, 1989). Phototrophs certainly power the system (Stephens, 1974; Lindsay et al., 2017), and anaerobic activities are prevalent (Boyd et al., 2017). The lake is a simple macro-ecosystem, birds eating two invertebrates, but the microbial communities and activities belie a high degree of complexity.

4.1 Bacteria, Archaea, and their viruses

Historically, GSL was often referred to as being sterile, "America's Dead Sea," (Scientific American, 1861; Deseret News, 1907) despite featuring brine that is teaming with life. We have known that GSL was not a sterile body of water for more than one hundred years as bacteria were first reported in the lake in the late nineteenth century (reviewed in Baxter, 2018). Early studies resulted in a few isolated species and reported 625 bacteria cells per milliliter of brine (Daniels [sic Daines, 1917). Claude Zobell and co-workers used a technique of submerging slides in GSL water, incubating in the lab, and counting colonies that grew on the slides (Zobell, 1937). A number of years later, Fred Post cultivated several bacterial and archaeal species from both the north and south arms, and he described them using the taxonomic framework existing at that time (Post, 1975, 1977, 1981). Post retired

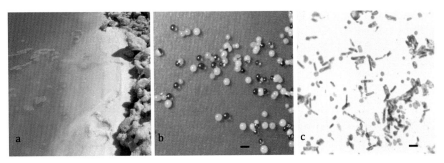

FIG. 4

(A) Pink north arm water at the shoreline; (B) halophilic archaeal and bacterial colonies on solid media, scale bar is 1 mm. (C) Light micrograph of a stained slide of north arm water, scale bar is 1 μm.

Image credits: Bonnie K. Baxter with (A Tabitha Webster, (B) Jason Rupp, and (C) Mike Acord.

before the Domain of archaea was understood and applied to taxonomy (Woese and Fox, 1977).

The pink color of the salt-saturated GSL north arm waters (Fig. 4A) reveals carotenoid-containing microorganisms in abundance. When this brine is plated on solid media in a petri dish, colorful colonies, in pink, purple, and orange, grow (Fig. 4B). Microscopic examination of stained preparations reveals the variety of cell morphologies of the bacteria and archaea present (Fig. 4C). These rosy-colored microorganisms in GSL were first reported by graduate students studying at the University of Utah (Daniels [sic Daines], 1917; Frederick, 1924; Kirkpatrick, 1934) and continue to engage scientists due to the photoprotection properties of these pigments (Jones and Baxter, 2017).

Building on the early studies, current knowledge of microbial life in GSL is based on traditional cultivation as well as non-cultivation (molecular) techniques. While culturing gives one lab strains with which one can explore physiology and biochemistry (e.g., Baxter et al., 2007; Pugin et al., 2012; D'Adamo et al., 2014), molecular biology gives us a more accurate system with which to construct taxonomic and relatedness data (e.g., Almeida-Dalmet et al., 2015; Ventosa et al., 2015). Researchers have cultivated a number of GSL strains of bacteria and archaea, isolating in laboratory conditions in salty media, to study their physiology and genetics. There are currently 13 GSL bacterial and one archaeal (*Methanohalophilus mahii*) strains stored and maintained in the following culture banks: the American Type Culture Collection (ATCC, USA), the Biological Resource Center (NBRC, Japan), the Leibniz Institute DSMZ (DSM, Germany), and the All-Russian Collection of Microorganisms (VKM, Russia). Table 2 presents these strains and information about the GSL site of isolation.

Assessing the DNA of an environment, through the 16SrRNA genes or from metagenomes, gives a more complete depiction of the community members since one captures the species that are not cultivatable. From such studies, we know that the

Table 2 GSL bacterial and archaeal strains available and stored in culture banks

Species	Strain designation	GSL isolation	Reference
Amphibacillus cookii	ATCC BAA-2118 DSM 23721 NBRC 11149	South arm sediment	Pugin et al. (2012)
Desulfobacter halotolerens	DSM 11383	South arm sediment	Brandt and Ingvorsen (1997)
Desulfocella halophile	ATCC 700426 DSM 11763	South arm sediment	Brandt et al. (1999)
Desulfohalobium utahense	DSM 17720 VKM B-2384	North arm water	Jakobsen et al. (2006)
Desulfosalsimonas propionicica	DSM 17721 VKM B-2385	North arm water	Kjeldsen et al. (2010)
Desulfovibrio sp.	DSM 12209	South arm sediment	Brandt et al. (1999)
Gracilibacillus halotolerans	ATCC 700849 DSM 11805	South arm sediment	Wainø et al. (1999)
Halanaerobium alcaliphilum	DSM 8275	South arm sediment	Tsai et al. (1995)
Halanaerobium praevalens	ATCC 33744 DSM 2228	South arm sediment	Zeikus et al. (1983)
Halobacillus trueperi	ATCC 700077 DSM 10404	South arm sediment	Spring et al. (1996)
Halobacillus litoralis	ATCC 700076 DSM 10405	South arm sediment	Spring et al. (1996)
Halomonas sp.	ATCC 27042	Solar evaporation pond	Gonzalez et al. (1972)
Halomonas utahensis	DSM 3051	North arm surface water	Fendrich (1988) and Sorokin and Tindall (2006)
Halomonas variabilis	ATCC 49240 DSM 3051 [III] NBRC 102410	South arm water	Fendrich (1988) and Dobson and Franzmann (1996)
Haloterrigena turkmenica	ATCC 51198 DSM 5511 JCM 9743 VKM B-1734	South arm sediment	Zvyagintseva and Tarasov (1987) and Ventosa et al. (1999)
Methanohalophilus mahii	ATCC 35705 DSM 5219	South arm sediment	Paterek and Smith (1985) and Paterek and Smith (1988)

microbial communities in GSL are composed predominantly of halophilic archaea and bacteria (Baxter et al., 2005; Weimer et al., 2009; Parnell et al., 2011; Meuser et al., 2013; Tazi et al., 2014; Almeida-Dalmet et al., 2015; Boogaerts, 2015). A search of "Great Salt Lake" in the GenBank database for deposited DNA sequences resulted in 862 hits for bacteria and 1230 hits for archaea (NLM, 2018). The lake is an unexplored reservoir of many uncultured and undescribed species; the majority of the 16S rRNA gene sequences in the GSL north arm were from uncultured taxa (Almeida-Dalmet et al., 2015). In general, the higher the salinity, the more archaeal genera are present relative to bacterial genera. (Eukaryotic algae, protozoa, and fungi are also present and will be discussed in more detail later.) These complex communities inhabit all the microniches of GSL such as the hypersaline north arm (Almeida-Dalmet et al., 2015) and the moderately saline south arm water (Meuser et al., 2013; Tazi et al., 2014). Also, bacteria and archaea inhabit the brine shrimp (Tkavc, 2012; Riddle et al., 2013) and their cysts (Baxter et al., 2007; Riddle et al., 2013).

These assemblages of microorganisms must be dynamic, responding to the changes in salinity and temperature that accompany the seasons in the high-altitude desert where they reside (Almeida-Dalmet et al., 2018). Salinity gradients in the less saline south arm of the lake have been shown to influence the composition of planktonic species composition (Boyd et al., 2014; Meuser et al., 2013; Lindsay et al., 2017). However, a temporal study of the hypersaline north arm microbiota demonstrated communities that are more stable over time and not as impacted by changes in temperature and salinity (Almeida-Dalmet et al., 2015). These stable hypersaline north arm microorganisms also have a lower phylogenetic diversity relative to communities in the south arm (Parnell et al., 2009, 2010, 2011).

The bacterial and archaeal population and profile composition are likely controlled by haloviruses, or halophage (Post, 1981; Baxter et al., 2011; Shen et al., 2012), which can shift the genera or species present based on specific virus-host interactions. Viral load in the north arm salt-saturated brine is 100:1, making them a significant group of predators in this extreme ecosystem (Baxter et al., 2011). The diversity of morphologies includes spherical, head-tail, fusiform (lemon-shaped), and filamentous structures (Fig. 5). Neither morphology diversity nor cell counts have been analyzed in the south arm, but plaques have been observed (Post, 1981) and a halophage was isolated with its bacterial host in this part of the lake (Shen et al., 2012). Clearly viruses play an important ecological role in GSL of controlling the microbial community profiles but also of gene exchange (Motlagh et al., 2017).

Metabolic activities by microorganisms, in general, occur more slowly or at lower levels as the salinity increases. For example, an examination of hydrocarbon and glutamate degradation in the evaporation ponds of a saltern at GSL noted a decrease in hydrocarbon turnover as salinity increased in the gradient of tested ponds (Ward and Brock, 1978). At moderate salinities, nitrogen cycling in south arm microcosms was stimulated by glutamate, but glutamate was degraded more slowly at higher salinities (Post and Stube, 1988), especially in north arm microcosms (Stube et al., 1976). Conversion of glucose, glycerol, and acetate to CO_2 was much slower in the North Arm than in the south arm (Fendrich and Schink, 1988).

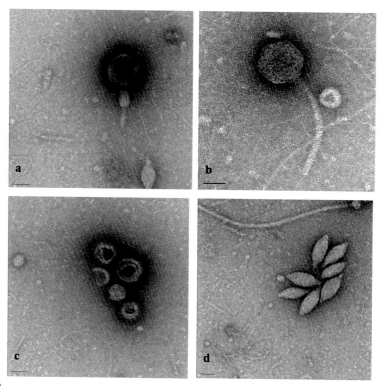

FIG. 5

Halovirus diversity in the north arm of Great Salt Lake, scale bar is 20 nm. (A) A head-tail virus with a two-stage tail; (B) a virus with a stiff, curved tail; (C) spherical viruses in a cluster; (D) A group of fusiform haloviruses, with a filamentous viral particle above them. Prepared as in Baxter et al. (2011).

Halovirus sampling, isolation TEM preparation and Image credit: Mihnea Mangalea, Bonnie K. Baxter, Smaranda Willcox, and Jack D. Griffith.

Regions of GSL are anoxic, whether in the sediments or in a stratified deep brine layer that sometime forms as higher density brine from the north arm seeps beneath the causeway into the south arm (Baxter et al., 2005; Naftz et al., 2009; Meuser et al., 2013). Thus anaerobic metabolisms such as methanogenesis or sulfate reduction can occur in these microniches. Biogenic methanogenesis was demonstrated in a mixed slurry of sediment from the north arm of GSL (Baxter et al., 2005), from south arm sediment samples (Phelps and Zeikus, 1980; Zeikus et al., 1983; Lupton et al., 1984), and from a microorganism, *Methanohalophilus mahii*, cultured from south arm sediment (Paterek and Smith, 1985; Paterek and Smith, 1988).

Several sulfate reducers have been isolated from GSL, including *Desulfobacter halotolerans* (Brandt and Ingvorsen, 1997) and *Desulfocella halophila* (Brandt et al., 1999). However, a large and diverse collection of sulfate-reducing bacteria from the north arm sediment was discovered in studies that employed molecular analyses (Kjeldsen et al., 2007; Boyd et al., 2017). Sulfate reduction activity with sediment samples decreased as salinity increased. In GSL, rates in various microniches vary widely, restricted by salinity and local anaerobic conditions (Brandt et al., 2001; Ingvorsen and Brandt, 2002; Boyd et al., 2017). Sulfate reduction chemistry favors mercury methylation by microorganisms, and given the high levels of mercury in both sediment and the deep brine layer, 80% of total mercury was found to be in the more toxic, organic, methyl-mercury form (Naftz et al., 2008). Wind and storm events (Beisnera et al., 2009 and Naftz et al., 2014) can bring methyl-mercury into the upper water column of GSL where it can enter the food web though the phytoplankton and into the birds (Johnson et al., 2015 and Jones and Wurtsbaugh, 2014).

4.2 Phytoplankton

Despite the extreme nature of this environment, GSL is highly productive and has a thriving photoautotrophic, or phytoplankton, community, which is controlled by environmental factors (Van Auken and McNulty, 1973; Marcarelli et al., 2006). Primary productivity by the phytoplankton in the south arm was reported to average 145 g C/m^2 per year in the south arm (Stephens and Gillespie, 1972; Stephens and Gillespie, 1976), oscillating seasonally dependent upon grazing by invertebrates (Hammer, 1981). A longitudinal study summarized nutrient data and pointed to sequestration when salinity stratification occurs, therefore nutrient availability and cycling is difficult to predict and measure (Belovsky et al., 2011). Inorganic nitrogen is considered to be the limiting factor for phytoplankton in the entire lake (Porcella and Holman, 1972; Wurtsbaugh, 1988). Nitrogen input during a five-year study was plentiful from atmospheric deposition and river input (Naftz, 2017), and the excess of nitrogen into the south arm, and its bays, has prompted concerns about eutrophication (Wurtsbaugh et al., 2009). The C:N ratio for the North Arm was determined to be 5.5:1 (Post, 1981). This isolated part of the lake does not have close proximity to an urban area like the south arm, but receives most of its nitrogen input from south arm flow through culverts or the breach (Naftz, 2017).

The phytoplankton are a key part of an aquatic ecosystem. They are the primary producers of GSL, and other saline systems, using photosynthesis to power the lake biota (Fig. 6A) (Hammer, 1981); however, these eukaryotes are typically four or more orders of magnitude lower in number than the bacteria and archaea in community profiles (Oren, 2014). The hypersalinity of the north arm limits phytoplankton, in fact, only a few species have been reported there (Post, 1977, 1980a; Felix and Rushforth, 1979; Lindsay et al., 2017). The south arm sports a broad diversity of phytoplankton, and the species richness increases with decreasing salinity (Felix and Rushforth, 1977, 1979, 1980; Rushforth and Felix, 1982; Stephens, 1990; Wurtsbaugh and Berry, 1990; Wurtsbaugh, 1992; Stephens, 1998; Larson and Belovsky, 2013). This means that environmental, temporal, and spatial factors that impact salinity can drive which species are represented in the community. Some species are excellent forage for the brine shrimp and brine fly larvae, while others are

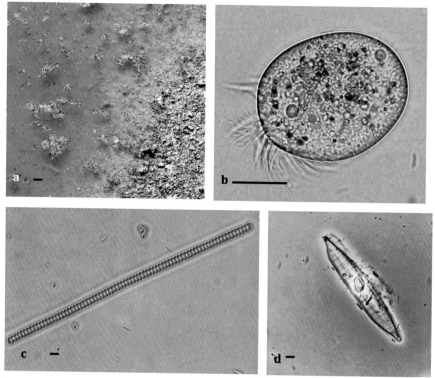

FIG. 6

Representative phytoplankton of Great Salt Lake. (A) Highly productive waters of the south arm, green due to the presence of phototrophic microorganisms, scale bar is 5 cm. (B) A green alga in the south arm water, scale bar is 5 μm. (C) An unidentified cyanobacteria from the south arm, scale bar is 5 μm. (D) A south arm diatom that is consistent in morphology with a *Navicula* species, scale bar is 5 μm.

Image credits: Isaac Hall and Bonnie K. Baxter.

not, and in this way, phytoplankton numbers and diversity can impact the entire ecosystem, limiting the invertebrates that feed the birds (Wurtsbaugh and Gliwicz, 2001; Belovsky et al., 2011).

The chlorophyta green algae are prevalent in the south arm, in fact all of the eukaryal 18S rRNA gene sequences in a recent study were affiliated with the related taxonomic order Chlamydomonadales (Meuser et al., 2013). Interesting early isolations include *Enteromorpha tubulosa* (Tilden, 1898) and *Tetraspora lubrica* (Flowers, 1934). Chlamydomonas was observed in early reports as well (Daniels [sic Daines], 1917), an observation recently supported with molecular studies (Meuser et al., 2013). In the more recent GSL ecosystem, *Dunaliella viridis* (Fig. 6B) and *Tetraselmis contracta* were overrepresented in the biomass in GSL microcosm experiments (Brock, 1975; Larson and Belovsky, 2013). In the lake waters of GSL, *Dunaliella* spp. dominate (Stephens and Gillespie, 1972; Stephens and Gillespie, 1976; Meuser et al., 2013). In a vertical transect study of south arm microbial communities, this genus was represented in all depth samples, from 71.4% to 96.4% of total sequences (Meuser et al., 2013). *Dunaliella* is especially prevalent during the highest primary productivity time of the year (the late winter to early spring) as the invertebrates in the water die off and grazing of algal cells stops (Stephens and Gillespie, 1976; Belovsky et al., 2011). During this time in the lake, the salinity has risen, following a hot summer and dry fall. *Dunaliella* spp. appear to flourish at lower temperatures and at these south arm higher salinities perhaps due to their ability to handle osmotic stress and thus outcompete other algae (Brock, 1975; Larson and Belovsky, 2013). The characteristic of osmophily in *Dunaliella* spp. is related to the production and accumulation of glycerol content inside their cells, balancing the salt solutes outside the cells (Oren, 2014).

The north arm phytoplankton is certainly less diverse. The more hypersaline-tolerant, *Dunaliella salina*, has been observed in the north arm saturated salinities above 300 g/L (Post, 1977, 1980a; Felix and Rushforth, 1979). This carotenoid-rich alga may survive salt saturation and desiccation by becoming encysted; the round cyst-like cells (aplanospores) of *D. salina* increased in representation in north arm microcosms at lower temperatures and higher salinity (Post, 1977). This alga can survive even under a salt crust and was deemed responsible for domes observed in the desiccated salty shore of the north arm in 1977, due to release of gases (Post, 1980b). More recently, a molecular analysis indicated the presence of another chlorophyte, *Tetracystis,* in the north arm (Lindsay et al., 2017).

Cyanobacteria are eubacteria (Fig. 6C), but due to the ability to photosynthesize, they are considered part of the phytoplankton community and are critical as food for the brine fly larvae (Aldrich, 1912; Wurtsbaugh, 2009). Several species were isolated and named in the earliest exploration of the GSL algae: *Polycystis packardii, Aphanothece utahensis, Dichothrix utahensis, Microcystis packardii,* and *Oscillatoria tenuis* (Tilden, 1898; Aldrich, 1912), but the salinity of sampling sites is not clear in these studies. *Coccochloris clabens* was isolated, also without specific location information (Flowers and Evans, 1966). Some cyanobacteria were likely isolated from brackish waters surrounding the lake and not the higher salinity open water (Stephens, 1974). In south arm water, four species of cyanobacteria were noted by microscopy (Felix and Rushforth, 1979), similar to the earlier observations and cultivation of five species, including two of *Chlamydomonas* (Kirkpatrick, 1934). Using molecular tools, a recent study identified the 16S rRNA genes of a *Euhalothece* species of halophilic cyanobacteria and suggests that this species may help form the microbialite structures of the lake (see Section 4.5) (Lindsay et al., 2017), which are grazed upon by brine fly larvae (Wurtsbaugh, 2009).

Diatoms of the GSL south arm were first noted by Daines (Daniels [sic Daines, 1917). Ruth Kirkpatrick tried to cultivate them, and she noted two species that resembled the *Navicula* genus (Kirkpatrick, 1934). Their presence was recently verified with genetic evidence of *Navicula* in the modern lake, associated with microbialites in the south arm (Lindsay et al., 2017). Microscopy of south arm water reveals a diversity of diatoms (e.g., Fig. 6D) (Hall and Baxter, unpublished). When they thrive in the south arm waters, competing out the chlorophytes, the brine shrimp population is negatively impacted, as the diatoms are likely a poor food source compared with *Dunaliella* (Belovsky et al., 2011).

In GSL, the various algal species have a niche in response to a salinity gradient, as exemplified by a south arm location with a less saline upper brine layer and the more saline deep brine layer (Meuser et al., 2013). For example, 18S rRNA gene sequences of *Oogamochlamys* spp. were observed in the vertical water column from the surface to 6 m deep, *Chlamydomonas* spp. genes were noted in a narrow range at 6.0 and 6.5 m depths, and *Picochlorum* genes were located at the 8 m depth. Salinity is clearly a driver of phytoplankton diversity.

4.3 Fungi

Both historical and modern studies on GSL ignored the possibility of fungi in the ecosystem. In fact, until recently, only sporadic reports existed regarding fungi in any natural hypersaline environment, one of them being the first report of a *Cladosporium* found on a submerged piece of pinewood from the GSL north arm (Cronin and Post, 1977). Past studies assumed that the growth of fungi at low water activity (a_w) reflected their general xerophilic/osmophilic behavior, as these fungi can grow at $a_w < 0.85$ (Pitt and Hocking, 1977; Pitt and Hocking, 1997). This corresponds to a medium supplemented with 17% NaCl or other solutes, like 50% glucose (Gunde-Cimerman et al., 2000, 2005). After isolation of numerous fungi from Mediterranean salterns (Gunde-Cimerman et al., 2000), it became clear that the growth of some species depends not only on the low a_w, but also on the chemical nature of the solutes that lower the a_w (de Hoog et al., 2005; Gunde-Cimerman et al., 2005, 2009).

The studies of fungi from natural saline systems was initiated in Sečovlje salterns (Slovenia), seasonal salterns at the Adriatic coast, and later continued at other salterns and salt lakes on three continents. Yeasts and filamentous fungi were isolated from GSL as well as the Dead Sea (Ein Bokek, Ein Gedi) and the Enriquillo Salt Lake (Dominican Republic) (Gunde-Cimerman et al., 2005). The black yeasts, which are extremely halotolerant species, *Hortaea werneckii*, *Phaeotheca triangularis*, and *Aureobasidium pullulans*, were commonly found in several of these locations, together with filamentous halotolerant and halophilic species of *Aspergillus* and *Penicillium* (Ascomycota) and halophilic genus *Wallemia* (Basidiomycota). Many yet unknown halotolerant species of the ubiquitous genus *Cladosporium* were described (Zalar et al., 2007). The occurrence of all these fungi has not only been documented in brine, but also on wood immersed in brine (Zalar et al., 2005), on the surface of halophytic plants (El-Morsy, 2000; Finkel et al., 2011; Ma et al., 2014), in saltern microbial mats (Cantrell et al., 2006; Tkavc, 2012), and even in brine ponds in deep seas (Burgaud et al., 2010). Extremely rare are the reports on chytrid fungi in hypersaline environments, but Amon (1978) reported on a sampling of four areas in and around the GSL, which revealed species of *Thraustochytrium*, *Schizochytrium*, *Labyrinthula*, and *Labyrinthuloides*. On the other hand, numerous yeasts are known from hypersaline environments (Zajc et al., 2017), *Debaryomyces hansenii* and *Metschnikowia bicuspidata* also from the less saline GSL south arm (Butinar et al., 2005), the presence of the latter was also confirmed by 18S rDNA sequences from a molecular study (Meuser et al., 2013).

We report here, for the first time, evidence for a rich diversity of fungi that live in the waters of GSL. In two GSL samplings performed in 2003 and 2009 on the samples of brine, salt, and oolites using the isolation methods described in Gunde-Cimerman et al. (2000), we isolated 32 fungal strains belonging to 11 genera (Table 3, Fig. 7), Ascomycota (genera *Acremonium, Alternaria, Aspergillus, Cladosporium, Coniochaeta, Neocamarosporium, Parengydontium, Penicillium, Stemphylium*) prevail the Basidiomycota, represented by a single genus (*Wallemia*). The fungal diversity in the less saline south arm is expectedly larger, the identified species already being reported in other studies from hypersaline environments.

Table 3 GSL fungal isolates preserved in the Microbial Culture Collection Ex, a cell bank for extremophiles, University of Ljubljana, Ljubljana, Slovenia (www.ex-genebank.com/)

Species	Strain designation (NCBI GenBank sequence accession numbers: *ITS rDNA, **beta-tubulin, ***actin, ****translation elongation factor 1-alpha)	GSL isolation site
Aspergillus fumigatus	EXF-1900 (MF189897**)	South arm
Cladosporium pseudocladosporioides	EXF-1901 (MF189933***)	South arm
Penicillium crustosum	EXF-1902 (MF189910*, MF189898**)	South arm
Cladosporium halotolerans	EXF-1903	South arm
Debaryomyces hansenii	EXF-1905 (MF189926*)	South arm
Cladosporium cladosporioides sp. complex	EXF-1908	South arm
Aspergillus flavus	EXF-1909 (MF189908*)	South arm
Acremonium af. sclerotigenum/ egyptiacum	EXF-1910 (MF189902*)	South arm
Wallemia sebi	EXF-1913 (MF189921*)	South arm
Cladosporium cladosporioides	EXF-1917 (MF189935***, MF189928****)	South arm
Cladosporium halotolerans	EXF-1918 (MF189936***, MF189929****)	South arm
Parengyodontium album	EXF-5591 (MF189925*)	North arm, Rozel Bay

Continued

Table 3 GSL fungal isolates preserved in the Microbial Culture Collection Ex, a cell bank for extremophiles, University of Ljubljana, Ljubljana, Slovenia (www. ex-genebank.com/) *Continued*

Species	Strain designation (NCBI GenBank sequence accession numbers: *ITS rDNA, **beta-tubulin, ***actin, ****translation elongation factor 1-alpha)	GSL isolation site
Penicillium buchwaldii	EXF-5592 (MF189911*)	North arm, Rozel Bay
Cladosporium herbarum sp. complex	EXF-5593 (MF189914*), EXF-5596 (MF189915*), EXF-5599 (MF189916*), EXF-5609 (MF189917*, MF189937***)	North arm, Rozel Bay
Alternaria multiformis	EXF-5595 (MF189906*)	North arm, Rozel Bay
Wallemia muriae	EXF-5597 (MF189922*)	North arm, Rozel Bay
Cladosporium cladosporioides sp. complex	EXF-5598 (MF189918*, MF189938***, MF189931****)	North arm, Rozel Bay
Alternaria rosae	EXF-5610 (MF189907*)	North arm, Rozel Bay
Neonectria punicea	EXF-5612 (MF189900**)	North arm, Rozel Bay
Acremonium sp.	EXF-5611 (MF189903*)	Pink halite sample from north arm
Stemphylium sp.	EXF-5613 (MF189927*)	Pink halite sample from north arm
Alternaria arborescens	EXF-5614 (MF189905*)	Pink halite sample from north arm
Cladosporium cladosporioides sp. complex	EXF-5600 (MF189919*), EXF-5601 (MF189920*, MF189939***, MF189932****)	Oolites from north arm shore
Penicillium aff. *mononematosum*	EXF-5602 (MF189912*, MF189901**)	Oolites from north arm shore
Alternaria arborescens	EXF-5616 (MF189904*)	Oolites from north arm shore
Neocamarosporium sp.	EXF-5618	Oolites from north arm shore
Aspergillus fumigatus	EXF-5622 (MF189909*)	Petroleum from north arm shore
Coniochaeta polymorpha	EXF-5624 (MF189924*)	Petroleum from north arm shore

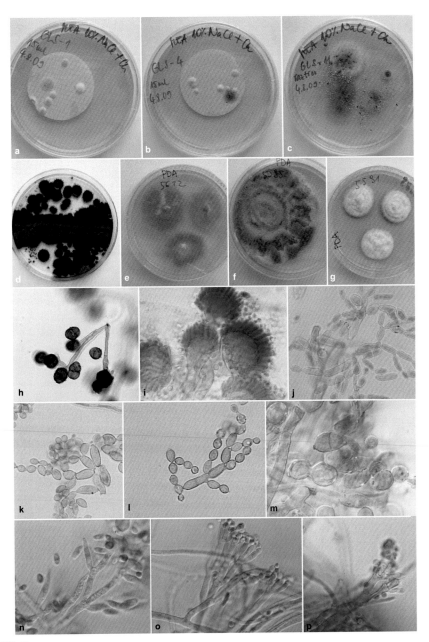

FIG. 7

Fungi from Great Salt Lake. (A) and (B) Primary isolation plates from brine. (C) Primary isolation plate from oolites. (D)–(G) Pure cultures of fungi: *Cladosporium* (D), *Neonectria* (E), *Neocamarosporium* (F), *Parengyodontium* (G). (H)–(P) Morphological structures under the light microscope of *Alternaria* (H), *Aspergillus* (I), *Cladosporium* (J–M), *Neonectria* (N), and *Penicillium* (O, P).

Among the isolates, strains belonging to the genus *Cladosporium* (Davidiella-ceae, Capnodiales) are the most numerous and several of them have not yet been identified to the species level, but presumably also represent some undescribed species. Since the strain of Cronin and Post (1977) was not preserved, and no DNA sequences are available, we cannot elucidate the species name. The taxonomy of the cosmopolitan genus *Cladosporium* has been revised (Bensch et al., 2012). The genus contains numerous species primarily inhabiting plants, as well as many salt extremotolerant species.

Representative strain of *Neocamarosporium* (Pleosporales) most likely belongs to *Neocamarosporium chichastianum*, but in culture it does not develop any sporulation structures, like typical pycnidial conidiomata for this species. *N. chichastianum* was described from soil of Lake Urmia in Iran (Crous et al., 2014), presumably saline. The GSL strain, *Coniochaeta polymorpha* (Coniochaetales), is for the first time reported here from a saline environment and it is known from a single clinical source (Khan et al., 2013). *Penicillium* species (Eurotiales), like *P. crustosum* and *P. buchwaldii* are known inhabitants of hypersaline brines, the later described as a potential producer of pharmacologically interesting compounds (Frisvad et al., 2013). The only species of *Aspergillus* (Eurotiales), *A. fumigatus*, is known from brines of salterns (Tepšič et al., 1997). The representatives related to here reported species *Penicillium mononematosum* were described from deserts and warm salt marshes soil in Egypt (Frisvad and Samson, 2004). The isolate from GSL oolites potentially represents a new, closely related species to *P. mononematosum*. The single isolate of *Neonectria* (Hypocreales), *N. punicea*, probably results from neighboring plants, since it is known from plant material (Lombard et al., 2013). Samples of pink salt from north arm revealed yet unknown taxa of *Acremonium* (Hypocreales) and *Stemphylium* (Pleosporales). In contrast to other hypersaline thalassohaline systems, no black yeasts have yet been found in GSL. The selective pressure of consistent high salinity may be too harsh for these organisms, which are typically found in areas with fluctuating salinity, from sea water salinity up to saturation. These preliminary studies provide the foundation to fuel future cultivation and molecular studies on the underexplored hypersaline fungi of GSL.

4.4 Protists

It was noted as early as 1917 that other single-celled eukaryotes (that were not fungi or phototrophs), the chemoheterotrophic protists, were also present in GSL (Vorhies, 1917). Consistent with this, protists are observed in other hypersaline systems (Hauer and Rogerson, 2005). As salinity changes, some protists may enter an encysted phase, which allows them to tolerate high salinities, then reemerge when the salinity decreases (Evans, 1958; Post et al., 1983). Although microscopy routinely reveals a diversity of protozoa in GSL, these remain an understudied part of the microbiota. Also, it is not certain which observed protists are salt-tolerant freshwater forms that find their way into the lake and survive but do not thrive (Evans, 1958).

One early study described observations of ciliates in the south arm brine, including *Uroleptus packii* (Pack, 1919). Other notable south arm isolations of ciliates include: *Pseudocohnilembus persalinus* (Evans and Thompson, 1964), *Euplotes parsalinus* (Evans, 1958, Evans and Thompson, 1964; Reddy, 1971), *Crystigera* sp., *Cyclidium* sp., and *Podophyra* sp. (Evans, 1958; Evans and Thompson, 1964). Two ciliates were isolated from north arm samples, a *Uronema* species (Stube et al., 1976), and *Chilophyra (Prorodon) utahensis* (Pack, 1919; Evans, 1958; Flowers and Evans, 1966).

Jones isolated a flagellate, *Euglena chamberlini*, and an amoeba, *Amoeba flowersi* from the south arm (Jones, 1944), followed by Evans who isolated an *Oikomonas* sp. (Evans, 1960). North arm hypersaline flagellates and amoeba were observed and described later in an aquarium experiment utilizing north arm sediment and water (Post, 1977). To date, no molecular analyses have been undertaken to identify the protozoa community of GSL.

4.5 Microbialites: microbial communities and geobiology

Microbialites are organosedimentary $CaCO_3$ deposits, which have accreted due to the actions of the benthic microbial community trapping and binding detrital sediment (Burne and Moore, 1987; Riding, 2000; Noffke and Awramik, 2013). Photosynthesis in aquatic environments removes CO_2 from the water, reducing the amount of carbonic acid present and increasing the pH (Riding, 2000; Dupraz and Visscher, 2005). This more alkaline local environment, in the presence of calcium from the groundwater, favors precipitation of $CaCO_3$ granules, which are bound and trapped by a microbial mat secreting sticky extracellular polymeric substance (Gebelein, 1969). The mat continues to photosynthesize, and the layers of intercalated mat and sediment grow, sometimes alternating in layers, causing lamination (e.g., stromatolite), or in a more disorganized manner with no lamination (e.g., thrombolite) (Kennard and James, 1986; Noffke and Awramik, 2013). Flowers (1934) suggested the occurrence of this process in GSL connected to the photosynthetic activity of cyanobacteria. GSL provides the perfect site for significant microbialite formation due to the shallow water, high solar penetration, high salinity, and unique geochemistry (Fig. 8).

The microbialites that line the lake bottom of GSL were first mapped by Eardley when the lake levels were low and the structures were visible (Eardley, 1938). More recent studies suggest the possibility that the microbialites form over faults, and that the faults in the GSL lakebed (Clark et al., in preparation; Dinter and Pechmann, 2014) are the source of groundwater (Colman et al., 2002; Vanden Berg et al., 2016), which could supply calcium. But there is only weak evidence for this at present linking groundwater to microbialite formation, and further work is needed. The GSL microbialite structures have varying morphologies depending on the local conditions of the site of formation (Carozzi, 1962). Much of the organosedimentary detrital sediment that composed the fabric of the GSL microbialites is oolitic sand

FIG. 8

Microbialites of Great Salt Lake (A) Vestige structures in the north arm. (B) Actively precipitating structures in the south arm.

Image credit: Bonnie K. Baxter.

(Chidsey et al., 2015). Cross-sections reveal variation among structures from the lake, some laminated and some disorganized.

Post was the first to examine the north arm carbonate structures and compare the biology of them to the south arm, showing the archaea-based community in the north versus a phototroph-rich community in the south arm (Post, 1977). Salinity gradients may affect the community structure as microbialites in GSL brine that was 160 g/L contained primarily cyanobacteria (Halley, 1976), but structures examined in water that was 130 g/L contained primarily diatoms (Collins, 1980). This was supported later with the observations of abundant chlorophyll associated with south arm microbialites and the presence of the cyanobacterial genus *Aphanothece* (Wurtsbaugh et al., 2011).

A recent study, with the goal of elucidating the microbial communities associated with the GSL microbialites, identified the likely "architects," cyanobacteria and diatoms, involved in their formation (Lindsay et al., 2017). Small subunit rRNA sequences from the phototrophs, *Euhalothece* and *Navicula,* were identified in abundance in the south arm microbialites. The south arm structures (Fig. 8B) showed features of active carbonate precipitation while the north arm structures (Fig. 8A) appear to be vestiges, which are inhabited only by "tenants," microorganisms that moved in after the architecture was completed. As the north arm was isolated and became hypersaline, the salinity tolerance of *Euhalothece* and *Navicula* was likely exceeded (Garcia-Pichel et al., 1998; Underwood and Provot, 2000). The only phototroph in the north arm structures, identified by18S rRNA gene sequence, was *Tetracystis*, a diatom, which has a high salinity tolerance (Kirst, 1989).

Microbialites clearly have an important role to play in the GSL food web. Their photosynthetic potential for harnessing solar energy is apparent in the high chlorophyll levels which are around nine times greater than in the surrounding brine (Wurtsbaugh et al., 2011). In addition, brine shrimp graze and brine fly pupate on the structures, which in turn impacts the distribution of migrating birds on the open water of GSL (Roberts and Conover, 2014). Microbialites form where groundwater seeps along fault lines, phytoplankton communities thrive, invertebrates feed, and birds thus congregate in these areas of rich food sources.

5 Great Salt Lake as an astrobiology analogue

GSL is a hypersaline ecosystem, a model extreme environment, and an analogue for salty sites on other space bodies in the universe. The halophilic microorganisms, or "halophiles," that thrive in the conditions described above, teach us about the limits of life on Earth, and these lessons may be applied elsewhere (Rothschild, 1990; Baxter et al., 2007, 2013).

5.1 Halophiles and the challenges for life in salt

To live in concentrated brine, halophiles must balance osmotically such that their cells do not shrivel up due to water loss. This is accomplished in part by the intracellular accumulation of osmotica, which balance against the salt on the outside of the cell membrane (Brown, 1976). Halophiles accumulate potassium ions and organic compatible solutes as their osmotica (Larsen, 1967; Galinski and Trüper, 1982; Galinski, 1993, 1995; Oren, 1999), which explains their success in salty environments like GSL. These extremophiles also have modifications in their proteins that help them function at high salt (Litchfield, 1998).

GSL halophilic microorganisms can survive desiccation in their environment (Baxter et al., 2007). In fact, such biota on Earth are very resistant to drought: endo-lithic halophilic cyanobacteria can survive in and around halite crystals in the hot, dry climate of the Atacama Desert (Wierzchos et al., 2006). Several studies point to the possibility of the survival of halophiles over geologic time scales in halite crystals (Norton and Grant, 1988; Norton et al., 1993; Denner et al., 1994; Grant et al., 1998; Stan-Lotter et al., 1999; McGenity et al., 2000; Vreeland et al., 2000; Stan-Lotter et al., 2002; Kminek et al., 2003; Mormile et al., 2003; Gruber et al., 2004; Park et al., 2009; Schubert et al., 2009, 2010; Lowenstein et al., 2011; Sankaranarayanan et al., 2011). Studies of modern halophiles that were desiccated and embedded in salt crystals indicate that the cells move toward the fluid inclusions and they can survive cold temperatures, but they suffer DNA damage (Fendrihan and Stan-Lotter, 2004). The mechanism of dormancy may involve low-grade metabolism and effective DNA repair processes (Johnson et al., 2007).

In addition to the high salt levels and desiccation, ultraviolet light exposure is intense in GSL and other hypersaline ecosystems. Halophiles are resistant to UV light (Dundas and Larsen, 1963; Baxter et al., 2007). When embedded in salt, they are particularly resistant (Fendrihan et al., 2009). They do have efficient DNA repair (McCready and Marcello, 2003; Kish and DiRuggiero, 2012; Jones and Baxter, 2017), but extreme halophilic microorganisms also typically contain carotenoid pigments in their membranes, which mitigate their photobiology by preventing DNA damage (White and Jahnke, 2002; Baxter et al., 2007). Halophilic archaea also possess genome sequence strategies to limit DNA damage such as bipyrimidine limitation (Jones and Baxter, 2016) and polyploidy (Jones and Baxter, 2017).

Clearly, halophiles can handle the osmotic challenges, the desiccation, and the high solar radiation of their environment. This makes halophiles multiextremophiles! All of these are potential challenges in space, and GSL organisms are critical to understanding life at its limits.

5.2 The preservation of biosignatures by halite and gypsum

Biomarkers in the rock record or biosignatures from preserved molecules are indicative of biological processes and are useful tools to probe ancient life forms in the paleontological record (Van den Kerkhof and Hein, 2001; Simoneit, 2004). On Earth, minerals may preserve labile biological constituents of cells over geologic time scales in highly protected microenvironments. Halite and gypsum in aqueous environments form crystalline structures with fluid inclusions inside (Roedder, 1984; Van den Kerkhof and Hein, 2001), in which biological materials can be stored within the mineral-saturated fluid, offering shelter from potentially damaging enzymatic, chemical, and physical processes and essentially serving as a time capsule (Griffith et al., 2008). Brine pockets, trapped in halite, have been used to study environmental parameters of the ancient environment in which they originated (Satterfield et al., 2005; Benison, 2006) and ancient microorganisms (e.g., Lowenstein et al., 2011). In astrobiology studies, important qualities of mineral fluid inclusions include photobiology concerns: these pockets must provide some protection from ultraviolet light radiation (Fendrihan et al., 2009), but they must transmit light from the part of the spectrum useful for photosynthesis (Rothschild, 1990; Cockell and Raven, 2004).

Molecules from this ancient fluid can provide clues about extinct life on early Earth or potentially on other space bodies. Which biological molecules can be preserved and for how long? Modern GSL studies in halite and gypsum will help us identify the parameters for preservation in minerals (Perl et al., 2016). Consideration should be made for the stability of each type of molecule and the environmental radiation exposure over time (Kminek et al., 2003; Fendrihan et al., 2009).

Nucleic acids, RNA and DNA, are obvious signatures for biology, and DNA is far more stable than RNA. Predictions of the rates of spontaneous DNA depurination, backbone cleavage, and other processes that degrade DNA suggest that large molecules should be degraded into small fragments within tens of thousands of years (Lindahl and Nyberg, 1972; Lindahl, 1993; Schroeder et al., 2006; Pääbo et al., 2004; Willerslev et al., 2004). In addition, these DNA molecules would lose some molecular integrity due to spontaneous chemical cross-linking (Hansen et al., 2006; Pääbo, 1989), which is critical in thinking about a detection strategy. However, DNA molecules were reportedly isolated from halite that is 425 million years old (Fish et al., 2002). Ancient gypsum inclusions have also been found to harbor DNA (Radax et al., 2001; Panieri et al., 2010). It is important to note then that the measurements of DNA degradation over time do not address DNA in saturated salts, where these reactions would occur more slowly. Also, there is evidence of DNA repair within these ancient cells (Johnson et al., 2007).

Carbohydrates may be preserved in halite. The polysaccharide, cellulose, is highly stable and relatively resistant to harsh conditions (Okuda et al., 1993; Kvien et al., 2005; Morán et al., 2008). Cellulose may be one of the most ancient biopolymers, and it is abundant within the biosphere of Earth (Cox et al., 2000). It is produced by cyanobacteria, perhaps one of the earliest known life forms, as well as other prokaryotic and eukaryotic species (Nobles et al., 2001). Cellulose can remain intact for a quarter of a billion years in halite fluid inclusions, as detected by electron microscopy (Griffith et al., 2008). Given that cellulose also has an infrared signature (Kondo and Sawatari, 1996), it could potentially be detectable on space bodies from orbit. This, taken with its stability over time, makes cellulose an important molecule for astrobiological studies.

Other biomolecules show promise of preservation as well. There are examples of protein preserved in the Earth environment, such as the 68 Ma collagen fragments isolated from dinosaur bones (Schweitzer et al., 2007). The building blocks of proteins, amino acids, may be key biomarkers; however, they can be produced by biotic or by abiotic processes (Bada, 2001). Methods have been developed to detect the amino acid chirality (biotic processes produce only the L-form while abiotic produce a 50:50 ratio of L and D forms), which can distinguish abiotic amino acids from those of biological sources (Hutt et al., 1999; Skelley and Mathies, 2003).

Lipid biomarkers as chemical fossils in the rock record can be used to indicate the types of microorganisms present and even the organization of life in aquatic microbial mats (e.g., Pawlowska et al., 2013). More significant here is the bio-preservation of the chemical structure of lipids over time, particularly in halite fluid inclusions (Winters, 2013) and in GSL recent sediment cores (Collister and Schamel, 2002). Cyclic and branched hydrocarbons are stable over geologic time (Simoneit, 2004). This group includes pigments such as carotenoid compounds, which are part of the halophile arsenal against high exposure to ultraviolet radiation (White and Jahnke, 2002; Baxter et al., 2007; Jones and Baxter, 2017).

Studies of ancient biomolecules and cells have been highly scrutinized, and technical problems including contamination have been noted (Pääbo et al., 2004; Hebsgaard et al., 2005). Of particular concern are analyses of environmental samples utilizing amplification methods such as the polymerase chain reaction (PCR) (e.g., Fish et al., 2002) or microbial cultivation techniques (e.g., Vreeland et al., 2000), which have the caveat of contamination possibilities. Other studies may avoid this with extensive surface sterilization (Sankaranarayanan et al., 2011) or by employing more direct methods like electron microscopy of the fluid from the inclusions, coupled with biochemical assays to identify the molecules (Griffith et al., 2008).

5.3 Great Salt Lake as an analogue for studying evaporites on Mars

On Earth, we see evidence of the minerals left behind when lakes dry up, and these evaporites are key to studying the ancient aquatic ecosystem that was once there (Sonnenfeld, 1984). GSL is surrounded by the Bonneville Salt Flats, a halite deposit left from the evaporation of ancient Lake Bonneville and other flooding/evaporation events (Turk, 1970). The GSL study site is a critical analogue as it provides a modern salt lake at the same location as an ancient evaporite. A study of the GSL system in the context of other Earthly salty environs gives us reference points for thinking about extreme salty life elsewhere in the Universe (Rothschild, 1990; Baxter et al., 2007).

The Salado Formation in New Mexico, USA, is an example of an older evaporite, 600 m of thick halite deposited 253 million years ago (Lowenstein, 1988; Holt and Powers, 1990; Powers et al., 2001; Beauheim and Roberts, 2002; Satterfield et al., 2005). Due to the overburden rock (Lowenstein, 1988), this halite deposit is protected from ionizing radiation and surface water contamination. This coupled with the fact that halite features low levels of naturally occurring radioactive isotopes means that the biological molecules in these halite fluid inclusions may be well preserved (Griffith et al., 2008). If we compare modern GSL and the surrounding 13,000-year-old Bonneville Salt Flats to this ancient system, we can envision time points in the process of evaporite formation and the preservation of the biological molecules that are sheltered there.

Ancient evaporites similar to halite deposits on Earth are found on the surface of Mars (Squyres et al., 2004), and halite and gypsum deposits have been described in the subsurface of the planet (Ehlmann et al., 2011). The Rover Opportunity discovered such mineral deposits, and thus evidence of ancient surface water, at Meridiani Planum, suggesting the area was once a lake that evaporated, concentrated the minerals, and left behind a thick salt crust (Squyres et al., 2004; Andrews-Hanna et al.,

2007). Like GSL, the mineral composition included abundant sulfates (Squyres et al., 2004). This suggests the existence of salt lakes on the surface of Mars, which could have been home to microbial life like the halophiles of GSL. Dissolved minerals in these bodies of water would have remained behind as the atmosphere dissipated over billions of years (Mancinelli et al., 2004). In addition, the halite deposits on Mars have are favorable to radiation that supports photosynthesis (Cockell and Raven, 2004). If life was there, it was most likely halophilic microbial life that could tolerate the water that became increasingly salty. These microorganisms could have managed with the osmotic stress as their ecosystem dried up.

In addition, Martian studies reveal the presence of modern salt water. The percolating groundwater at this Meridiani Planum is both salty and acidic (Tosca et al., 2008). At another location, spectroscopy analysis from the Mars Reconnaissance Orbiter revealed hydrated salts associated with contemporary flowing water on Mars (Ojha et al., 2015). Present-day active brine on the surface of Mars, as well as evaporites from ancient salt water, points the search for life toward halophilic microorganisms.

In saturated brine, water molecules interacting with ions are less available to support life, and some have theorized that life cannot tolerate the saturated Martian acidic brines (Tosca et al., 2008). However, we have evidence on Earth of halophilic microorganisms that do just that, despite the low water activity (Fendrihan et al., 2012), even in acidic salt lakes (Benison et al., 2008). In fact, strong comparisons can be made to ancient saline lake deposits on Earth with those on Mars (Benison, 2006). The Earth's salt lakes, including the more neutral or basic ones, such as GSL, remain powerful analogues to understanding life in high salt, one of the more likely possibilities for Martian life.

Acknowledgments

The authors would like to thank all the researchers who have worked on Great Salt Lake for the last century and a half, many referenced here. For thoughtful editing and scientific input, we are grateful to Jaimi Butler, Kendall Tate, Mike Dyall-Smith, Nina Gunde-Cimerman, Jack Oviatt, Don L. Clark, Mike Vanden Berg, and John Neill. A number of the photo credits go to our students whose eyes and minds are opened by this amazing ecosystem. Funding for many of our lake explorations at Westminster College has come from the W.M. Keck Foundation and the Lawrence T. and Janet T. Dee Foundation.

References

Aldrich, J.M., 1912. The biology of some western species of the dipterous genus *Ephydra*. J. N. Y. Entomol. Soc. 20, 77–102.

Aldrich, T.W., Paul, D.S., 2002. Avian ecology of Great Salt Lake. In: Gwynn, J.W. (Ed.), Great Salt Lake: An Overview of Change. Utah Geological Survey, State of Utah Department of Natural Resources, Salt Lake City, UT, pp. 343–374.

Almeida-Dalmet, S., Sikaroodi, M., Gillevet, P.M., Litchfield, C.D., Baxter, B.K., 2015. Temporal study of the microbial diversity of the North Arm of Great Salt Lake, Utah, US. Microorganisms 3 (3), 310–326.

Almeida-Dalmet, S., Litchfield, C.D., Gillevet, P., Baxter, B.K., 2018. Differential gene expression in response to salinity and temperature in a *Haloarcula* strain from great salt Lake, Utah. Gene 9 (1), 52.

Amon, J.P., 1978. Thraustochytrids and Labyrinthulids of terrestrial, aquatic and hypersaline environments of the Great Salt Lake, USA. Mycologia 70 (6), 1299–1301.

Andrews-Hanna, J.C., Phillips, R.J., Zuber, M.T., 2007. Meridiani Planum and the global hydrology of Mars. Nature 446 (7132), 163–166.

Atwood, G., Wambeam, T.J., Anderson, N.J., 2016. The present as a key to the past: paleoshoreline correlation insights from Great Salt Lake. In: Oviatt, C.G., Shroder, J.F. (Eds.), Lake Bonneville a Scientific Update. Elsevier, Netherlands.

Bada, J.L., 2001. State-of-the-art instruments for detecting extraterrestrial life. Proc. Natl. Acad. Sci. 98 (3), 797–800.

Barnes, B.D., Wurtsbaugh, W.A., 2015. The effects of salinity on plankton and benthic communities in the Great Salt Lake, Utah, USA: a microcosm experiment. Can. J. Fish. Aquat. Sci. 72 (6), 807–817.

Baxter, B.K., 2018. Great Salt Lake microbiology: a historical perspective. Int. Microbiol, 1–17.

Baxter, B.K., Litchfield, C.D., Sowers, K., Griffith, J.D., Dassarma, P.A., Dassarma, S., 2005. Microbial diversity of Great Salt Lake. Microbial diversity of Great Salt Lake. In: Gunde-Cimerman, N., Oren, A., Plemenitaš, A. (Eds.), Adaptation to Life at High Salt Concentrations in Archaea, Bacteria, and Eukarya. Cellular Origin, Life in Extreme Habitats and Astrobiology. In: vol. 9. Springer, Dordrecht, Netherlands, pp. 9–25.

Baxter, B.K., Eddington, B., Riddle, M.R., Webster, T.N., Avery, B.J., 2007. Great Salt Lake halophilic microorganisms as models for astrobiology: evidence for desiccation tolerance and ultraviolet radiation resistance. In: Hoover, R.B., Levin, G.V., Rozanov, A.Y., Davies, P.C.W. (Eds.), Instruments, Methods, and Missions for Astrobiology X. In: vol. 6694. SPIE, Bellingham, WA, p. 669415.

Baxter, B.K., Mangalea, M.R., Willcox, S., Sabet, S., Nagoulat, M.N., Griffith, J.G., 2011. Haloviruses of Great Salt Lake: a model for understanding viral diversity. In: Ventosa, A., Oren, A., Ma, Y. (Eds.), Halophiles and Hypersaline Environments: Current Research and Future Trends. Springer, Dordrecht, Netherlands.

Baxter, B.K., Butler, J.K., Kleba, B., 2013. Worth Your Salt: halophiles in education. In: Vreeland, R.H. (Ed.), Advances in Understanding the Biology of Halophilic Microorganisms. Springer, Dordrecht, Netherlands.

Beauheim, R.L., Roberts, R.M., 2002. Hydrology and hydraulic properties of a bedded evaporite formation. J. Hydrol. 259, 66–88.

Behrens, P., 1980. Industrial processing of Great Salt Lake brines by Great Salt Lake minerals and chemicals corporation. In: Gwynn, J.W. (Ed.), Great Salt Lake: A Scientific, Historical and Economic Overview. Utah Geological Survey, State of Utah Department of Natural Resources, Salt Lake City, UT, pp. 223–228.

Beisnera, K., Naftz, D.L., Johnson, W.P., Diaz, X., 2009. Selenium and trace element mobility affected by periodic displacement of stratification in the Great Salt Lake, Utah. Sci. Total Environ. 407 (19), 5263–5273.

Bellrose, F.C., 1980. Ducks, Geese, and Swans of North America. Stackpole Books, Harrisburg, Pennsylvania, USA.

Belovsky, G.E., Stephens, D., Perschon, C., Birdsey, P., Paul, D., Naftz, D., Baskin, R., Larson, C., Mellison, C., Luft, J., Mosley, R., 2011. The Great Salt Lake ecosystem (Utah, USA): long term data and a structural equation approach. Ecosphere 2 (3), 1–40.

Benison, K.C., 2006. A Martian analog in Kansas: comparing Martian strata with Permian acid saline lake deposits. Geology 34, 385–388.

Benison, K.C., Jagniecki, E.A., Edwards, T.B., Mormile, M.R., Storrie-Lombardi, M.C., 2008. "Hairy blobs:" microbial suspects preserved in modern and ancient extremely acid lake evaporites. Astrobiology 8 (4), 807–821.

Bensch, K., Braun, U., Groenewald, J.Z., Crous, P.W., 2012. The genus *Cladosporium*. Stud. Mycol. 72, 1–401.

Bingham, C.P., 1980. Solar production of potash from the brines of the Bonneville salt flats. In: Gwynn, J.W. (Ed.), Great Salt Lake: A Scientific, Historical and Economic Overview. Utah Geological Survey, State of Utah Department of Natural Resources, Salt Lake City, UT, pp. 229–242.

Boogaerts, G.L., 2015. Preliminary Characterization of the Microbial Community in the Bonneville Salt Flats. (Thesis). The University of Alabama, Birmingham.

Boyd, E.S., Hamilton, T.L., Swanson, K.D., Howells, A.E., Baxter, B.K., Meuser, J.E., Posewitz, M.C., Peters, J.W., 2014. [FeFe]-hydrogenase abundance and diversity along a vertical redox gradient in Great Salt Lake, USA. Int. J. Mol. Sci. 15, 21947–21966.

Boyd, E.S., Yu, R.-Q., Barkay, T., Hamilton, T.L., Baxter, B.K., Naftz, D.L., Marvin-DiPasquale, M., 2017. Effect of salinity on mercury methylating benthic microbes and their activities in Great Salt Lake, Utah. Sci. Total Environ. 581–582, 495–506. https://doi.org/10.1016/j.scitotenv.2016.12.157.

Brandt, K.K., Ingvorsen, K., 1997. *Desulfobacter halotolerans* sp. nov., a halotolerant acetate-oxidizing sulfate-reducing bacterium isolated from sediments of Great Salt Lake, Utah. Syst. Appl. Microbiol. 20 (3), 366–373.

Brandt, K.K., Patel, B.K., Ingvorsen, K., 1999. *Desulfocella halophila* gen. nov., sp. nov., a halophilic, fatty-acid-oxidizing, sulfate-reducing bacterium isolated from sediments of the Great Salt Lake. Int. J. Syst. Evol. Microbiol. 49, 193–200.

Brandt, K.K., Vester, F., Jensen, A.N., Ingvorsen, K., 2001. Sulfate reduction dynamics and enumeration of sulfate-reducing bacteria in hypersaline sediments of the Great Salt Lake (Utah, USA). Microb. Ecol. 41 (1), 1–11.

Brock, T.D., 1975. Salinity and the ecology of *Dunaliella* from Great Salt Lake. Microbiology 89, 285–292. https://doi.org/10.1099/00221287-89-2-285.

Brown, A.D., 1976. Microbial water stress. Bacteriol. Rev. 40, 803–846.

Burgaud, G., Arzur, D., Durand, L., Cambon-Bonavita, M.A., Barbier, G., 2010. Marine culturable yeasts in deep-sea hydrothermal vents: diversity and association with fauna. FEMS Microbiol. Ecol. 73 (1), 121–133. https://doi.org/10.1111/j.1574-6941.2010.00881.x.

Burne, R.V., Moore, L.S., 1987. Microbialites: organosedimentary deposits of benthic microbial communities. PALAIOS 2 (3), 241–254.

Butinar, L., Santos, S., Spencer-Martins, I., Oren, A., Gunde-Cimerman, N., 2005. Yeast diversity in hypersaline habitats. FEMS Microbiol. Lett. 244 (2), 229–234. https://doi.org/10.1016/j.femsle.2005.01.043.

Cannon, J.S., Cannon, M.A., 2002. The southern Pacific railroad trestle—past and present. In: Gwynn, J.W. (Ed.), Great Salt Lake: An Overview of Change. Utah Geological Survey, State of Utah Department of Natural Resources, Salt Lake City, UT, pp. 283–294.

Cantrell, S.A., Casillas-MartiNez, L., Molina, M., 2006. Characterization of fungi from hypersaline environments of solar salterns using morphological and molecular techniques. Mycol. Res. 110 (8), 962–970. https://doi.org/10.1016/j.mycres.2006.06.005.

Carozzi, A.V., 1962. Observations on algal biostromes in the Great Salt Lake, Utah. J. Geol. 70, 246–252.

Chidsey Jr., T.C., Vanden Berg, M.D., Eby, D.E., 2015. Petrography and characterization of microbial carbonates and associated facies from modern Great Salt Lake and Uinta Basin's Eocene Green River formation in Utah, USA. Microb. Carbon. Space Time: Implications Global Explor. Prod. 418, 261–286.

Clark, D.L., Oviatt, C.G. & Dinter, D.A., in preparation, Interim geologic map of the Tooele 30′ x 60′ quadrangle, Tooele, Salt Lake, and Davis Counties, Utah, year 4: Utah Geological Survey Open-File Report, GIS data, scale 1:62,500.

Cockell, C.S., Raven, J.A., 2004. Zones of photosynthetic potential on Mars and the early Earth. Icarus 169 (2), 300–310.

Cohenour, R.E., Thompson, K.C., 1966. Geologic Setting of Great Salt Lake.

Collins, N., 1980. Population ecology of *Ephydra cinerea* Jones (Diptera: Ephydridae), the only benthic metazoan of the great salt Lake, USA. Hydrobiologia 68, 99–112.

Collister, J.W., Schamel, S., 2002. Lipid composition of recent sediments from the Great Salt Lake. In: Gwynn, J.W. (Ed.), Great Salt Lake: An Overview of Change. Special Publication of Utah Department of Natural Resources, Salt Lake City, UT, pp. 127–142.

Colman, S.M., Kelts, K.R., Dinter, D.A., 2002. Depositional history and neotectonics in Great Salt Lake, Utah, from high-resolution seismic stratigraphy. Sediment. Geol. 148 (1), 61–78.

Cook, B.I., Ault, T.R., Smerdon, J.E., 2015. Unprecedented 21st century drought risk in the American Southwest and Central Plains. Sci. Adv. 1 (1), e1400082.

Cox, P.M., Betts, R.A., Jones, C.D., Spall, S.A., Totterdell, I.J., 2000. Acceleration of global warming due to carbon-cycle feedbacks in a coupled climate model. Nature 408, 184–187.

Cronin, E.A., Post, F.J., 1977. Report of a dematiaceous hyphomycete from the Great Satl Lake, Utah. Mycologia 69, 846–847.

Crosman, E.T., Horel, J.D., 2009. Modis-derived surface temperature of the Great Salt Lake. Remote Sens. Environ. 113, 73–81.

Crous, P.W., Groenewald, J.Z., Papizadeh, M., Abolhassan, S., Fazeli, S., Amoozegar, M.A., 2014. *Neocamarosporium chichastianum* Papizadeh, Crous, Shahzadeh Fazeli, Amoozegar, sp. nov. Persoonia: Mol. Phylogeny Evol. Fungi 293 (33), 236–237. https://doi.org/10.3767/003158514X685680.

D'Adamo, S., Jinkerson, R.E., Boyd, E.S., Brown, S.L., Baxter, B.K., Peters, J.W., Posewitz, M., 2014. Evolutionary and biotechnological implications of robust hydrogenase activity in halophilic strains of *Tetraselmis*. PLoS ONE. 9 (1): e85812, https://doi.org/10.1371/journal.pone.0085812.

Damsté, J.S.S., De Leeuw, J.W., Kock-Van Dalen, A.C., De Zeeuw, M.A., De Lange, F., Irene, W., Rijpstra, C., Schenck, P.A., 1987. The occurrence and identification of series of organic sulphur compounds in oils and sediment extracts. I. A study of Rozel point oil (U.S.A.). Geochim. Cosmochim. Acta 51, 2369–2391.

Daniels [sic Daines], L.L., 1917. On the flora of Great Salt Lake. Am. Nat. 51, 499–506.

de Hoog, G.S., Zalar, P., van den Ende, B.G., Gunde-Cimerman, N., 2005. Relation of halotolerance to human pathogenicity in the fungal tree of life: an overview of ecology and evolution under stress. In: Gunde-Cimerman, N., Oren, A., Plemenitas, A. (Eds.),

Adaptation to Life at High Salt-Concentration in Archaea, Bacteria and Eukarya. In: vol. 9. Dordrecht, Netherlands, Springer, pp. 373–395.

Deamer, K., 2016. Utah's Great Salt Lake is Shrinking. http://www.livescience.com/57055-utah-great-salt-lake-shrinking.html/. (accessed 07.12.18).

Denner, E.B.M., McGenity, T.J., Busse, H.-J., Grant, W.D., Wanner, G., Stan-Lotter, H., 1994. *Halococcus salifodinae* sp. nov., an archaeal isolate from an Austrian salt mine. Int. J. Syst. Evol. Microbiol. 44 (2), 774–780.

Deseret Evening News, 1907. Weary Path Troden by Intrepid Band to the Shores of America's Dead Sea. Deseret Evening News, Salt Lake City, Utah.

Diaz, X., Johnson, W.P., Oliver, W., Naftz, D.L., 2009. Volatile selenium flux from the Great Salt Lake, Utah. Environ. Sci. Technol. 43, 53–59.

Dinter, D.A. & Pechmann, J.C. (2014). Paleoseismology of the Promontory Segment, East Great Salt Lake Fault. *Final Technical Report for U.S. Geological Survey*, award number 02HQGR0105, 23 p.

Dobson, S.J., Franzmann, P.D., 1996. Unification of the genera *Deleya* (Baumann et al. 1983), *Halomonas* (Vreeland et al. 1980), and *Halovibrio* (Fendrich 1988) and the species *Paracoccus halodenitrificans* (Robinson and Gibbons 1952) into a single genus, *Halomonas*, and placement of the genus *Zymobacter* in the family Halomonadaceae. *Int*. J. System. Bacteriol. 46, 550–558.

Domagalski, J.L., Orem, W.H., Eugster, H.P., 1989. Organic geochemistry and brine composition in Great Salt, Mono, and Walker Lakes. Geochim. Cosmochim. Acta 53 (11), 2857–2872.

Domagalski, J.L., Eugster, H.P., Jones, B.F., 1990. Trace metal geochemistry of Walker, Mono, and Great Salt Lakes. In: Spencer, R.J., Chou, , I-Ming (Eds.), Fluid-Minerals Interactions: A Tribute to H.P. Eugster. Geochemical Society, Washington, D.C., pp. 315–353.

Dundas, I.D., Larsen, H., 1963. A study on the killing by light of photosensitized cells of *Halobacterium salinarium*. Arch. Microbiol. 46, 19–28.

Dupraz, C., Visscher, P.T., 2005. Microbial lithification in marine stromatolites and hypersaline mats. Trends Microbiol. 13, 429–438.

Eardley, A.J., 1938. Sediments of Great Salt Lake, Utah. Am. Assoc. Prof. Geol. Bull. 22 (10), 1305–1411.

Eardley, A.J., Stringham, B., 1952. Selenite crystals in the clays of Great Salt Lake. J. Sediment. Petrol. 22 (4), 234–238.

Ehlmann, B.L., Mustard, J.F., Murchie, S.L., Bibring, J.P., Meunier, A., Fraeman, A.A., Langevin, Y., 2011. Subsurface water and clay mineral formation during the early history of Mars. Nature 479 (7371), 53–60.

EI-Morsy, E.M., 2000. Fungi isolated from the endorhizosphere of halophytic plants from the Red Sea coast of Egypt. Fungal Divers. 5, 43–54.

Evans, F.R., 1958. Culture of protozoa from Great Salt Lake. J. Protozool. 5, 13.

Evans, F.R., 1960. Studies on growth of protozoa from the Great Salt Lake with special reference to *Cristigera* sp. J. Protozool. 7, 14–15.

Evans, F.R., Thompson, J.C., 1964. *Pseudocohnilembidae* n. fam., a hymenostome ciliate family containing one genus, *Pseudocohnilembus* ng, with three new species. J. Protozool. 11 (3), 344–352.

Felix, E.A., Rushforth, S.R., 1977. The algal flora of the Great Salt Lake, Utah: a preliminary report. In: Greer, D. (Ed.), Desertic Terminal Lakes. Utah Water Resources Lab, Logan, Utah, pp. 385–392.

Felix, E.A., Rushforth, S.R., 1979. The algal flora of the Great-Salt-Lake, Utah, USA. Nova Hedwigia 31, 163–195.

Felix, E.A., Rushforth, S.R., 1980. Biology of the south arm of the Great Salt Lake, Utah. Utah Geol. Mineral. Surv. Bull. 116, 305–312.

Fendrich, C., 1988. *Halovibrio variabilis* gen. nov.sp. nov., *Pseudomonas halophila* sp. nov. and a new halophilic aerobic coccoid eubacterium from Great Salt Lake, Utah, USA. Syst. Appl. Microbiol. 11, 36–43.

Fendrich, C., Schink, B., 1988. Degradation of glucose, glycerol, and acetate by aerobic bacteria in surface water of Great Salt Lake, Utah, U.S.A. Syst. Appl. Microbiol. 11, 94–96.

Fendrihan, S., Stan-Lotter, H., 2004. Survival of halobacteria in fluid inclusions as a model of possible biotic survival in Martian halite. Mars Planet. Sci. Technol, 9–18.

Fendrihan, S., Bérces, A., Lammer, H., Musso, M., Rontó, G., Polacsek, T.K., Holzinger, A., Kolb, C., Stan-Lotter, H., 2009. Investigating the effects of simulated Martian ultraviolet radiation on *Halococcus dombrowskii* and other extremely halophilic archaebacteria. Astrobiology 9 (1), 104–112.

Fendrihan, S., Dornmayr-Pfaffenhuemer, M., Gerbl, F.W., Holzinger, A., Grösbacher, M., Briza, Erler, P., Gruber, C., Plätzer, K., Stan-Lotter, H., 2012. Spherical particles of halophilic archaea correlate with exposure to low water activity—implications for microbial survival in fluid inclusions of ancient halite. Geobiology 10 (5), 424–433.

Finkel, O.M., Burch, A.Y., Lindow, S.E., Post, A.F., Belkin, S., 2011. Geographical location determines the population structure in phyllosphere microbial communities of a salt-excreting desert tree. Appl. Environ. Microbiol. 77, 7647–7655.

Fish, S.A., Shepherd, T.J., McGenity, T.J., Grant, W.D., 2002. Recovery of 16S ribosomal RNA gene fragments from ancient halite. Nature 420, 432–436.

Flowers, S., 1934. Vegetation of the Great Salt Lake region. Bot. Gaz. 95, 353–418.

Flowers, S., Evans, F.K., 1966. The flora and fauna of the Great Salt Lake region, Utah. In: - Boyko, H. (Ed.), Salinity and Aridity. New Approaches to Old Problems. Dr. W. Junk Publishers, The Hague, Netherlands, pp. 367–393.

Frederick, E., 1924. On the Bacterial Flora of Great Salt Lake and the Viability of Other Microorganisms in Great Salt Lake Water (Master's thesis). University of Utah. 65 pp.

Frisvad, J.C., Samson, R.A., 2004. Polyphasic taxonomy of *Penicillium* subgenus *Penicillium*. A guide to identification of food and air-borne terverticillate Penicillia and their mycotoxins. Stud. Mycol. 49, 1–174.

Frisvad, J.C., Houbraken, J., Popma, S., Samson, R.A., 2013. Two new *Penicillium* species *Penicillium buchwaldii* and *Penicillium spathulatum*, producing the anticancer compound asperphenamate. FEMS Microbiol. Lett. 339, 77–92.

Fritz, A., 2014. Great Salt Lake Approaches 167 Year Record Low. Washington Post. https://www.washingtonpost.com/news/capital-weather-gang/wp/2014/08/19/great-salt-lake-approaches-167-year-record-low/?utm_term=.d0330351547e/. (accessed 9.08.18).

Galinski, E.A., 1993. Compatible solutes of halophilic eubacteria: molecular principles, water-solute interaction, stress protection. Experientia 49 (6–7), 487–496.

Galinski, E.A., 1995. Osmoadaptation in bacteria. Adv. Microb. Physiol. 37, 273–328.

Galinski, E.A., Trüper, H.G., 1982. Betaine, a compatible solute in the extremely halophilic phototrophic bacterium *Ectothiorhodospira halochloris*. FEMS Microbiol. Lett. 13 (4), 357–360.

Garcia-Pichel, F., Nubel, U., Muyzer, G., 1998. The phylogeny of unicellular, extremely halotolerant cyanobacteria. Arch. Microbiol. 169, 469–482.

Gebelein, C.D., 1969. Distribution morphology and accretion rate of recent subtidal algal stromatolites, Bermuda. J. Sediment. Petrol. 39, 49–69.

Gonzalez, C.F., Taber, W.A., Zeitoun, M.A., 1972. Biodegradation of ethylene glycol by a salt-requiring bacterium. Appl. Microbiol. 24, 911–919.

Grant, W.D., Gemmell, R.T., McGenity, T.J., 1998. Halobacteria: the evidence for longevity. Extremophiles 2 (3), 279–287.

Greer, D.C., 1971. Annals map supplement fourteen: Great Salt Lake, Utah. Ann. Assoc. Am. Geogr. 61, 214–215.

Griffith, J.D., Willcox, S., Powers, D.W., Nelson, R., Baxter, B.K., 2008. Discovery of abundant cellulose microfibers encased in 250 Ma Permian halite: a macromolecular target in the search for life on other planets. Astrobiology 8 (2), 215–228.

Gruber, C., Legat, A., Pfaffenhuemer, M., Radax, C., Weidler, G., Busse, H.J., Stan-Lotter, H., 2004. *Halobacterium noricense* sp. nov., an archaeal isolate from a bore core of an alpine Permian salt deposit, classification of *Halobacterium* sp. NRC-1 as a strain of *H. salinarum* and emended description of *H. salinarum*. Extremophiles 8 (6), 431–439.

Gunde-Cimerman, N., Zalar, P., De Hoog, S., Plemenitaš, A., 2000. Hypersaline waters in salterns—natural ecological niches for halophilic black yeasts. FEMS Microb. Ecol. 32, 235–240.

Gunde-Cimerman, N., Butinar, L., Sonjak, S., Turk, M., Uršič, V., Zalar, P., Plemenitaš, A., 2005. Halotolerant and halophilic fungi from coastal environments in the Arctic. In: - Gunde-Cimerman, N., Oren, A., Plemenitaš, A. (Eds.), Adaptation to Life at High Salt Concentrations in Archaea, Bacteria, and Eukarya. Cellular Origin, Life in Extreme Habitats and Astrobiology. In: vol. 9. Springer, Dordrecht, Netherlands, pp. 397–423.

Gunde-Cimerman, N., Ramos, J., Plemenitaš, A., 2009. Halotolerant and halophilic fungi. Mycol. Res. 113, 1231–1241.

Gwynn, J.W., 1998. Great Salt Lake, Utah: chemical and physical variations of the brine and effects of the SPRR causeway, 1966–1996. In: Modern and Ancient Lake Systems: New Problems and Perspectives. AAPG, Tulsa, OK, pp. 71–90.

Halley, R.B., 1976. Textural variation within Great Salt Lake algal mounds. Dev. Sedimentol. 20, 435–445.

Halley, R.B., 1977. Ooid fabric and fracture in the Great Salt Lake and the geologic record. J. Sediment. Petrol. 47 (3), 1099–1120.

Hammer, U.T., 1981. Primary production in saline lakes. Hydrobiologia 81, 47–57. https://doi.org/10.1007/BF00048705.

Hansen, A.J., Mitchell, D.L., Wiuf, C., Paniker, L., Brand, T.B., Binladen, J., Gilichinsky, D.A., Rønn, R., Willerslev, E., 2006. Crosslinks rather than strand breaks determine access to ancient DNA sequences from frozen sediments. Genetics 173 (2), 1175–1179.

Hauer, G., Rogerson, A., 2005. Heterotrophic protozoa from hypersaline environments. In: Gunde-Cimerman, N., Oren, A., Plemenitaš, A. (Eds.), Adaptation to Life at High Salt Concentrations in Archaea, Bacteria, and Eukarya. Cellular Origin, Life in Extreme Habitats and Astrobiology. In: vol. 9. Springer, Dordrecht, Netherlands, pp. 519–539.

Hebsgaard, M.B., Phillips, M., Willerslev, E., 2005. Geologically ancient DNA: fact or artefact? Trends Microbiol. 13, 212–220.

Holt, R.M., Powers, D.W., 1990. Geological and hydrological studies of evaporites in the northern Delaware Basin for the waste isolation pilot plant (WIPP), New Mexico: guidebook 14. In: Powers, D.W., Holt, R.M., Beauheim, R.L., Rempe, N. (Eds.), Geological Society of America Annual Meeting. Dallas Geological Society, Dallas, Texas, pp. 45–78.

Hutt, L.D., Glavin, D.P., Bada, J.L., Mathies, R.A., 1999. Microfabricated capillary electrophoresis amino acid chirality analyzer for extraterrestrial exploration. Anal. Chem. 71 (18), 4000–4006.

Ingvorsen, K., Brandt, K.K., 2002. Anaerobic microbiology and sulfur cycling in hypersaline sediments with special reference to Great Salt Lake. In: Gwynn, J.W. (Ed.), Great Salt Lake: An Overview of Change. Utah Department of Natural Resources, Salt Lake City, Utah, pp. 385–396.

Jakobsen, T., Kjeldsen, K., Ingvorsen, K., 2006. *Desulfohalobium utahense* sp. nov., a moderately halophilic, sulfate-reducing bacterium isolated from Great Salt Lake. Int. J. Syst. Evol. Microbiol. 56 (9), 2063–2069. https://doi.org/10.1099/ijs.0.64323-0.

Javor, B., 1989. Hypersaline Environments, Microbiology and Biogeochemistry. Springer-Verlag, Berlin.

Johnson, S.S., Hebsgaard, M.B., Christensen, T.R., Mastepanov, M., Nielsen, R., Munch, K., Rønn, R., 2007. Ancient bacteria show evidence of DNA repair. Proc. Natl. Acad. Sci. 104 (36), 14401–14405.

Johnson, W.P., Swanson, N., Black, B., Rudd, A., Carling, G., Fernandez, D.P., Luft, J., Van Leeuwen, J., Marvin-DiPasquale, M., 2015. Total-and methyl-mercury concentrations and methylation rates across the freshwater to hypersaline continuum of the Great Salt Lake, Utah, USA. Sci. Total Environ. 511, 489–500.

Jones, D.T., 1944. Two protozoans from Great Salt Lake. Bull. Univ. Utah 35 (8), 1–11.

Jones, D.L., Baxter, B.K., 2016. Bipyrimidine signatures as a photoprotective genome strategy in G+ C-rich halophilic archaea. Life 6 (3), 37.

Jones, D.L., Baxter, B.K., 2017. DNA repair and photoprotection: mechanisms of overcoming environmental ultraviolet radiation exposure in halophilic archaea. In: Recent Advances in DNA Repair. Frontiers in Microbiology. vol. 8, p. 1882.

Jones, E.F., Wurtsbaugh, W.A., 2014. The Great Salt Lake's monimolimnion and its importance for mercury bioaccumulation in brine shrimp (*Artemia franciscana*). Limnol. Oceanogr. 59 (1), 141–155.

Jones, B.F., Naftz, D.L., Spencer, R.J., Oviatt, C.G., 2009. Geochemical evolution of Great Salt Lake, Utah, USA. Aquat. Geochem. 15 (1–2), 95–121.

Keck, W., Hassibe, W., 1979. The Great Salt Lake. U.S. Geol. Surv, 25.

Kennard, J.M., James, N.P., 1986. Thrombolites and stromatolites: two distinct types of microbial structures. PALAIOS 1 (5), 492–503.

Khan, Z., Gené, J., Ahmad, S., Cano, J., Al-Sweih, N., Joseph, L., Chandy, R., Guarro, J., 2013. *Coniochaeta polymorpha*, a new species from endotracheal aspirate of a preterm neonate, and transfer of *Lecythophora* species to *Coniochaeta*. Antonie Van Leeuwenhoek 104, 243–252.

Kirkpatrick, R., 1934. The Life of Great Salt Lake, With Special Reference to the Algae (Master's thesis). University of Utah. 30 pp.

Kirst, G.O., 1989. Salinity tolerance of eukaryotic marine algae. Annu. Rev. Plant Physiol. Plant Mol. Biol. 40, 21–53.

Kish, A., DiRuggiero, J., 2012. DNA replication and repair in halophiles. In: Vreeland, R.H. (Ed.), Advances in Understanding the Biology of Halophilic Microorganisms. Springer, Dordrecht, Netherlands, pp. 163–198.

Kjeldsen, K.U., Loy, A., Jakobsen, T.F., Thomsen, T.R., Wagner, M., Ingvorsen, K., 2007. Diversity of sulfate-reducing bacteria from an extreme hypersaline sediment, Great Salt Lake (Utah). FEMS Microbiol. Ecol. 60 (2), 287–298. https://doi.org/10.1111/j.1574-6941.2007.00288.x.

Kjeldsen, K.U., Jakobsen, T.F., Glastrup, J., Ingvorsen, K., 2010. Desulfosalsimonas propionicica gen. nov., sp. nov., a halophilic, sulfate-reducing member of the family Desulfobacteraceae isolated from a salt-lake sediment. Int. J. Syst. Evol. Microbiol. 60, 1060–1065. https://doi.org/10.1099/ijs.0.014746-0.

Kminek, G., Bada, J.L., Pogliano, K., Ward, J.F., 2003. Radiation-dependent limit for the viability of bacterial spores in halite fluid inclusions and on Mars. Radiat. Res. 159, 722–729.

Kondo, T., Sawatari, C., 1996. A Fourier transform infra-red spectroscopic analysis of the character of hydrogen bonds in amorphous cellulose. Polymer 37 (3), 393–399.

Kvien, I., Tanem, B.S., Oksman, K., 2005. Characterization of cellulose whiskers and their nanocomposites by atomic force and electron microscopy. Biomacromolecules 6, 3160–3165.

Larsen, H., 1967. Biochemical aspects of extreme halophilism. Adv. Microb. Physiol. 1, 97–132.

Larson, C.A., Belovsky, G.E., 2013. Salinity and nutrients influence species richness and evenness of phytoplankton communities in microcosm experiments from Great Salt Lake, Utah, USA. J. Plankton Res. 35, 1154–1166. https://doi.org/10.1093/plankt/fbt053.

Lindahl, T., 1993. Instability and decay of the primary structure of DNA. Nature 362 (6422), 709–715.

Lindahl, T., Nyberg, B., 1972. Rate of depurination of native deoxyribonucleic acid. Biochemistry 11 (19), 3610–3618.

Lindsay, M.R., Anderson, C., Fox, N., Scofield, G., Allen, J., Anderson, E., Bueter, L., Poudel, S., Sutherland, K., Munson-McGee, J.H., Van Nostrand, J.D., Zhou, J., Spear, J.R., Baxter, B.K., Lageson, D.R., Boyd, E.S., 2017. Microbialite response to an anthropogenic salinity gradient in Great Salt Lake, Utah. Geobiology 15 (1), 131–145.

Litchfield, C.D., 1998. Survival strategies for microorganisms in hypersaline environments and their relevance to life on early Mars. Meteorit. Planet. Sci. 33 (4), 813–819.

Lombard, L., van der Merwe, N.A., Groenewald, J.Z., Crous, P.W., 2013. Lineages in Nectriaceae: re-evaluating the generic status of Ilyonectria and allied genera. Phytopathol. Mediterr. 53, 340–357.

Lowenstein, T.K., 1988. Origin of depositional cycles in a Permian "saline giant": the Salado (McNutt zone) evaporites of New Mexico and Texas. Geol. Soc. Am. Bull. 100, 592–608.

Lowenstein, T.K., Schubert, B.A., Timofeeff, M.N., 2011. Microbial communities in fluid inclusions and long-term survival in halite. Geol. Soc. Am. Today 21 (1), 4–9.

Lupton, F.S., Phelps, T.J., Zeikus, J.G., 1984. Methanogenesis, sulphate reduction and hydrogen metabolism in hypersaline anoxic sediments of the Great Salt Lake, Utah. In: Annual Report, Baas Becking Geobiological Laboratory. Bureau of Mineral Resources, Canberra, Australia, pp. 42–48.

Ma, Y., Zhang, H., Du, Y., Tian, T., Xiang, T., Liu, X., Wu, F., An, L., Wang, W., Gu, J.-D., Feng, H., 2014. The community distribution of bacteria and fungi on ancient wall paintings of the Mogao Grottoes. Sci. Rep. 5, 7752. https://doi.org/10.1038/srep07752.

Madison, R.J., 1970. Effects of a causeway on the chemistry of the brine in Great Salt Lake Utah. Water-Resour. Bull. 14.

Mancinelli, R.L., Fahlen, T.F., Landheim, R., Klovstad, M.R., 2004. Brines and evaporites: analogs for Martian life. Adv. Space Res. 33 (8), 1244–1246.

Marcarelli, A.M., Wurtsbaugh, W.A., Griset, O., 2006. Salinity controls phytoplankton response to nutrient enrichment in the Great Salt Lake, Utah, USA. Can. J. Fish. Aquat. Sci. 63, 2236–2248.

McCready, S., Marcello, L., 2003. Repair of UV damage in Halobacterium salinarum. Biochem. Soc. Trans. 31 (3), 694–698. https://doi.org/10.1042/bst0310694.

McGenity, T.J., Gemmell, R.T., Grant, W.D., Stan-Lotter, H., 2000. Origins of halophilic microorganisms in ancient salt deposits. Environ. Microbiol. 2 (3), 243–250.

Meuser, J.E., Baxter, B.K., Spear, J.R., Peters, J.W., Posewitz, M.C., Boyd, E.S., 2013. Contrasting patterns of community assembly in the stratified water column of Great Salt Lake, Utah. Microb. Ecol. 66 (2), 268–280.

Morán, J.I., Alvarez, V.A., Cyras, V.P., Vázquez, A., 2008. Extraction of cellulose and preparation of nanocellulose from sisal fibers. Cellulose 15 (1), 149–159.

Mormile, M.R., Biesen, M.A., Gutierrez, M.C., Ventosa, A., Pavlovich, J.B., Onstott, T.C., Fredrickson, J.K., 2003. Isolation of *Halobacterium salinarum* retrieved directly from halite brine inclusions. Environ. Microbiol. 5 (11), 1094–1102.

Motlagh, A.M., Bhattacharjee, A.S., Coutinho, F.H., Dutilh, B.E., Casjens, S.R., Goel, R.K., 2017. Insights of phage-host interaction in hypersaline ecosystem through metagenomics analyses. Front. Microbiol. 8, 352.

Naftz, D.L., 2017. Inputs and internal cycling of nitrogen to a causeway influenced, hypersaline Lake, Great Salt lake, Utah, USA. Aquat. Geochem. 23 (3), 199–216.

Naftz, D.L., Angeroth, C., Kenney, T., Waddell, B., Darnall, N., Silva, S., Perschon, C., Whitehead, J., 2008. Anthropogenic influences on the input and biogeochemical cycling of nutrients and mercury in Great Salt Lake, Utah, USA. Appl. Geochem. 23 (6), 1731–1744.

Naftz, D.L., Fuller, C., Cederberg, J., Krabbenhoft, D., Whitehead, J., 2009. Mercury inputs to Great Salt Lake, Utah: reconnaissance-phase results. Natl. Resour. Environ. Issues 15 (5), 37–49.

Naftz, D.L., Millero, F.J., Jones, B.F., Green, W.R., 2011. An equation of state for hypersaline water in Great Salt Lake, Utah, USA. Aquat. Geochem. 17, 809–820.

Naftz, D.L., Carling, G.T., Angeroth, C., Freeman, M., Rowland, R., Pazmiño, E., 2014. Density-stratified flow events in Great Salt Lake, Utah, USA: implications for mercury and salinity cycling. Aquat. Geochem. 20 (6), 547–571.

National Library of Medicine, 2018. National Center for Biotechnology Information. https://www.ncbi.nlm.nih.gov/nuccore/?term=%22great+salt+lake%22/. (accessed 09.03.18).

Neill, J., Leite, B., Gonzales, J., Sanchez, K., Luft, J., 2016. 2015 Great Salt Lake Eared Grebe Aerial Photo Survey. Annual Report, Great Salt Lake Ecosystem Program. State of Utah Division of Wildlife Resources, Salt Lake City, UT.

Nobles, D.R., Romanovicz, D.K., Brown Jr., R.M., 2001. Cellulose in cyanobacteria: origin of vascular plant cellulose synthase? Plant Physiol. 127, 529–542.

Noffke, N., Awramik, S., 2013. Stromatolites and MISS—differences between relatives. Geol. Soc. Am. Today 23 (9), 4–9.

Norton, C.F., Grant, W.D., 1988. Survival of Halobacteria within fluid inclusions in salt crystals. J. Gen. Microbiol. 134, 1365–1373.

Norton, C.F., McGenity, T.J., Grant, W.D., 1993. Archaeal halophiles (halobacteria) from two British salt mines. Microbiology 139 (5), 1077–1081.

Ojha, L., Wilhelm, M.B., Murchie, S.L., McEwen, A.S., Wray, J.J., Hanley, J., Massé, M., Chojnacki, M., 2015. Spectral evidence for hydrated salts in recurring slope lineae on Mars. Nat. Geosci. 8 (11), 829–832.

Okuda, K., Kudlicka, K., Kuga, S., Brown Jr., R.M., 1993. [beta]-Glucan synthesis in the cotton fiber (I. identification of [beta]-1,4- and [beta]-1,3-glucans synthesized in vitro). Plant Physiol. 101, 1131–1142.

Oren, A., 1993. Ecology of extremely halophilic microorganisms. In: Vreeland, R.H., Hochstein, L.I. (Eds.), The Biology of Halophilic Bacteria. CRC Press, Boca Raton, FL, pp. 25–53.

Oren, A., 1999. Bioenergetic aspects of halophilism. Microbiol. Mol. Biol. Rev. 63 (2), 334–348.

Oren, A., 2014. The ecology of *Dunaliella* in high-salt environments. J. Biol. Res.-Thessaloniki 21 (1), 23. https://doi.org/10.1186/s40709-014-0023-y.

Oring, L.W., Neel, L., Oring, K.E., 2000. Intermountain West Regional Shorebird Plan, Version 1.0. 48 pp. https://www.shorebirdplan.org/wpcontent/uploads/2013/01/IMWEST4.pdf. (Accessed 9 March 2018).

Oviatt, C.G., Thompson, R.S., Kaufman, D.S., Bright, J., Forester, R.M., 1999. Reinterpretation of the Burmester Core, Bonneville Basin, Utah. Quat. Res. 52, 180–184.

Pääbo, S. (1989). Ancient DNA: extraction, characterization, molecular cloning, and enzymatic amplification. Proc. Natl. Acad. Sci. U. S. A. (1939–1943). National Academy of Sciences, Washington, DC.

Pääbo, S., Poinar, H., Serre, D., Jaenicke-Després, V., Hebler, J., Rohland, N., Kuch, M., Krause, J., Vigilant, L., Hofreiter, M., 2004. Genetic analyses from ancient DNA. Annu. Rev. Genet. 38, 645–679.

Pack, D., 1919. Two Ciliata of Great Salt Lake. Biol. Bull. 36 (4), 273–282.

Packard Jr., A.S., 1871. On insects inhabiting salt water. Am. J. Sci. 3 (1), 100–110.

Panieri, G., Lugli, S., Manzi, V., Roveri, M., Schreiber, B.C., Palinska, K.A., 2010. Ribosomal RNA gene fragments from fossilized cyanobacteria identified in primary gypsum from the late Miocene, Italy. Geobiology 8 (2), 101–111.

Park, J.S., Vreeland, R.H., Cho, B.C., Lowenstein, T.K., Timofeeff, M.N., Rosenzweig, W.D., 2009. Haloarchaeal diversity in 23, 121 and 419 MYA salts. Geobiology 7 (5), 515–523.

Parnell, J.J., Crowl, T.A., Weimer, B.C., Pfrender, M.E., 2009. Bio-diversity in microbial communities: system scale patterns and mechanisms. Mol. Ecol. 18, 1455–1462.

Parnell, J.J., Rompato, G., Latta, L.C., Pfrender, M.E., Van Nostrand, J.D., He, Z., Zhou, J., Andersen, G., Champine, P., Ganesan, B., Weimer, B.C., 2010. Functional biogeography as evidence of gene transfer in hypersaline microbial communities. PLoS ONE 5 (9) e12919.

Parnell, J.J., Rompato, G., Crowl, T.A., Weimer, B.C., Pfrender, M.E., 2011. Phylogenetic distance in Great Salt Lake microbial communities. Aquat. Microb. Ecol. 64, 267–273.

Paterek, J.R., Smith, P.H., 1985. Isolaton and characterization of a halophilic methanogen from Great Salt Lake. Appl. Environ. Microbiol. 50, 877–881.

Paterek, J.R., Smith, P.H., 1988. Methanohalophilus *mahii* gen. nov., sp. nov., a methylotrophic halophilic methanogen. Int. J. Syst. Bacteriol. 38, 122–123.

Paul, D.S., Manning, A.E., 2016. Great Salt Lake Waterbird Survey Five-Year Report (1997–2001). http://www.wildlife.utah.gov/gsl/waterbirdsurvey/. (Accessed 5 May 2018).

Pawlowska, M.M., Butterfield, N.J., Brocks, J.J., 2013. Lipid taphonomy in the Proterozoic and the effect of microbial mats on biomarker preservation. Geology 41 (2), 103–106.

Perl, S.M., Vaishampayan, P.A., Corsetti, F.A., Piazza, O., Ah-med, M., Willis, P., Creamer, J.S., Williford, K.W., Flannery, D.T., Tuite, M.L., Ehlmann, B.L., Bhartia, R., Baxter, B.K., Butler, J.K., Hodyss, R., Berelson, W.M., Nealson, K.H., 2016. Identification and validation of bigenic preservation: defining contraints within Martian mineralogy. Proceedings of the Biosignature Preservation and Detection in Mars Analog Environments Conference, Lake Tahoe, NV.

Peterson, C., Gustin, M., 2008. Mercury in the air, water and biota at the Great Salt Lake (Utah, USA). Sci. Total Environ. 405 (1–3), 255–268.

Phelps, T., Zeikus, J.G., 1980. Microbial ecology of anaerobic decomposition in Great Salt Lake. In: Proceedings of the Annual Meeting of the American Society for Microbiology, Abstract 14.p. 89.

Pitt, J.I., Hocking, A.D., 1977. Influence of solute and hydrogen-ion concentration on water relations of some xerophilic fungi. J. Gen. Microbiol. 101, 35–40.

Pitt, J.I., Hocking, A.D., 1997. Fungi and Food Spoilage. Blackie Academic & Professional, London, New York.

Porcella, D.B., Holman, J.A., 1972. Nutrients, algal growth, and culture of brine shrimp in the southern Great Salt Lake. In: Riley, J.P. (Ed.), Proceedings of The Great Salt Lake and Utah's water resources: First Annual Conference of Utah Section American Water Research Association. Utah State University, Utah Water Research Laboratory, pp. 142–155.

Post, F.J., 1975. Life in the Great Salt Lake, Utah. Science 36, 43–47.

Post, F.J., 1977. The microbial ecology of the Great Salt Lake. Microb. Ecol. 3, 143–165.

Post, F.J., 1980a. Biology of the north arm. In: Gwynn, J.W. (Ed.), Great Salt Lake: An Overview of Change. State of Utah Department of Natural Resources, Salt Lake City, UT, pp. 314–321.

Post, F.J., 1980b. Oxygen-rich gas domes of microbial origin in the salt crust of the Great Salt Lake, Utah. Geomicrobiol J. 2, 127–139. https://doi.org/10.1080/01490458009377757.

Post, F.J., 1981. Microbiology of the Great Salt Lake North Arm. Hydrobiologia 81, 59–69.

Post, F.J., Stube, J.C., 1988. A microcosm study of nitrogen utilization in the Great Salt Lake, Utah. Hydrobiolgia 158, 89–100.

Post, F.J., Borowitzka, L.J., Borowitzka, M.A., Mackay, B., Moulton, T., 1983. The protozoa of a Western Australian hypersaline lagoon. Hydrobiologia 105 (1), 95–113. https://doi.org/10.1007/BF00025180.

Powers, D.W., Vreeland, R.H., Rosenzweig, W.D., 2001. How old are bacteria from the Permian age? Nature 411, 155–156.

Pugin, B., Blamey, J.M., Baxter, B.K., Wiegel, J., 2012. *Amphibacillus cookii* sp. nov., a facultatively aerobic, spore-forming, moderately halophilic, alkalithermotolerant bacterium. Int. J. Syst. Evol. Microbiol. 62 (9), 2090–2096.

Radax, C., Gruber, C., Stan-Lotter, H., 2001. Novel haloarchaeal 16S rRNA gene sequences from Alpine Permo-Triassic rock salt. Extremophiles 5, 221–228.

Reddy, Y.J.R., 1971. A Description of a New Species of *Euplotes* From Great Salt Lake, Utah (Unpublished master's thesis). University of Utah, Salt Lake City, Utah.

Riddle, M.R., Baxter, B.K., Avery, B.J., 2013. Molecular identification of microorganisms associated with the brine shrimp, *Artemia franciscana*. Aquatic Biosyst. 9, 1–11.

Riding, R., 2000. Microbial carbonates: the geological record of calcified bacterial-algal mats and biofilms. Sedimentology 47, 179–214.

Roberts, A.J., 2013. Avian diets in a saline ecosystem: Great Salt Lake, Utah, USA. Human–Wildlife Interact. 7, 158–168.

Roberts, A.J., Conover, M.R., 2014. Diet and body mass of ducks in the presence of commercial harvest of brine shrimp cysts in the Great Salt Lake, Utah. J. Wildl. Manag. 78 (7), 1197–1205.

Roedder, E., 1984. The fluids in salt. Am. Mineral. 69, 413–439.

Rothschild, L.J., 1990. Earth analogs for Martian life. Icarus 88 (1), 246–260. https://doi.org/10.1016/0019-1035(90)90188-F.

Rupke, A.L., McDonald, A., 2012. Great Salt Lake Brine Chemistry Database, 1966–2011. Utah Geological Survey, State of Utah Department of Natural Resources, Salt Lake City, UT.

Rushforth, S.R., Felix, E.A., 1982. Biotic adjustments to changing salinities in the Great Salt Lake, Utah, USA. Microb. Ecol. 8, 157–161.

Sandberg, P.A., 1975. New interpretations of Great Salt Lake ooids and of ancient non-skeletal carbonate mineralogy. Sedimentology 22, 497–537. https://doi.org/10.1111/j.1365-3091.1975.tb00244.x.

Sankaranarayanan, K., Timofeeff, M.N., Spathis, R., Lowenstein, T.K., Lum, J.K., 2011. Ancient microbes from halite fluid inclusions: optimized surface sterilization and DNA extraction. PLoS ONE 6. (6): e20683, https://doi.org/10.1371/journal.pone.0020683.

Satterfield, C.L., Lowenstein, T.K., Vreeland, R.H., Rosenzweig, W.D., 2005. Paleobrine temperatures, chemistries, and paleoenvironments of Silurian Salina Formation F-1 salt, Michigan Basin, U.S.A., from petrography and fluid inclusions in halite. J. Sediment. Res. 75, 534–546.

Saxton, H.J., Goodman, J.R., Collins, J.N., Black, F.J., 2013. Maternal transfer of inorganic mercury and methylmercury in aquatic and terrestrial arthropods. Environ. Toxicol. Chem. 3, 2630–2636. https://doi.org/10.1002/etc.2350.

Schroeder, G.K., Lad, C., Wyman, P., Williams, N.H., Wolfenden, R., 2006. The time required for water attack at the phosphorus atom of simple phosphodiesters and of DNA. Proc. Natl. Acad. Sci. U. S. A. 103 (11), 4052–4055.

Schubert, B.A., Lowenstein, T.K., Timofeeff, M.N., 2009. Microscopic identification of prokaryotes in modern and ancient halite, Saline Valley and Death Valley, California. Astrobiology 9 (5), 467–482.

Schubert, B.A., Lowenstein, T.K., Timofeeff, M.N., Parker, M.A., 2010. Halophilic Archaea cultured from ancient halite, Death Valley, California. Environ. Microbiol. 12 (2), 440–454.

Schweitzer, M.H., Suo, Z., Avci, R., Asara, J.M., Allen, M.A., Arce, F.T., Horner, J.R., 2007. Analyses of soft tissue from *Tyrannosaurus rex* suggest the presence of protein. Science 316 (5822), 277–280.

Scientific American, 1861. Great Salt Lake. 9, pp. 131–132.

Shen, P.S., Domek, M.J., Sanz-García, E., Makaju, A., Taylor, R.M., Hoggan, R., Culumber, M.D., Oberg, C.J., Breakwell, D.P., Prince, J.T., Belnap, D.M., 2012. Sequence and structural characterization of Great Salt Lake Bacteriophage CW02, a member of the T7-like supergroup. J. Virol. 86 (15), 7907–7917. https://doi.org/10.1128/JVI.00407-12.

Shroder, J.F., Cornwell, K., Oviatt, C.G., Lowndes, T.C., 2016. Landslides, Alluvial Fans, and Dam Failure at Red Rock Pass: the outlet of Lake Bonneville. In: Oviatt, C.G., Shroder, J.F. (Eds.), Lake Bonneville a Scientific Update. Elsevier, Netherlands.

Simoneit, B.R., 2004. Biomarkers (molecular fossils) as geochemical indicators of life. Adv. Space Res. 33 (8), 1255–1261.

Sinninghe-Damsté, J.S., De Leeuw, J.W., Kock-Van Dalen, A.C., De Zeeuw, M.A., De Lange, F., Rijpstra, W.I.C., Schenck, P.A., 1987. The occurrence and identification of series of organic sulphur compounds in oils and sediment extracts. I. A study of Rozel Point Oil (U.S.A.). Geochim. Cosmochim. Acta 51, 2369–2391.

Skelley, A.M., Mathies, R.A., 2003. Chiral separation of fluorescamine-labeled amino acids using microfabricated capillary electrophoresis devices for extraterrestrial exploration. J. Chromatogr. A 1021 (1), 191–199.

Sonnenfeld, P., 1984. Brines and Evaporites. University of California, Academic Press, Oakland, CA.

Sorokin, D.Y., Tindall, B.J., 2006. The status of the genus name *Halovibrio* Fendrich 1989 and the identity of the strains *Pseudomonas halophila* DSM 3050 and *Halomonas variabilis* DSM 3051. Request for an opinion. Int. J. Syst. Evol. Microbiol. 56 (2), 487–489.

Spencer, R.J., Eugster, H.P., Jones, B.F., Rettig, S.L., 1985. Geochemistry of Great Salt Lake, Utah I: hydrochemistry since 1850. Geochim. Cosmochim. Acta 49 (3), 727–737.

Spring, S., Ludwig, W., Marquez, M.C., Ventosa, A., Schleifer, K.H., 1996. *Halobacillus* gen. nov., with descriptions of *Halobacillus litoralis* sp. nov. and *Halobacillus trueperi* sp. nov., and transfer of *Sporosarcina halophila* to *Halobacillus halophilus* comb. nov. Int. J. Syst. Bacteriol. 46, 492–496.

Squyres, S.W., Grotzinger, J.P., Arvidson, R.E., Bell, J.F., Calvin, W., Christensen, P.R., Clark, B.C., Crisp, J.A., Farrand, W.H., Herkenhoff, K.E., Johnson, J.R., 2004. In situ evidence for an ancient aqueous environment at Meridiani Planum. Mars. Science 306 (5702), 709–1714.

Stan-Lotter, H., McGenity, T.J., Legat, A., Denner, E.B., Glaser, K., Stetter, K.O., Wanner, G., 1999. Very similar strains of Halococcus salifodinae are found in geographically separated Permo-Triassic salt deposits. Microbiology 145 (12), 3565–3574.

Stan-Lotter, H., Pfaffenhuemer, M., Legat, A., Busse, H.-J., Radax, C., Gruber, C., 2002. *Halococcus dombrowskii* sp. nov., an archaeal isolate from a Permo-Triassic alpine salt deposit. Int. J. Syst. Bacteriol. 52, 1807–1814.

Stansbury, H., 1855. Exploration of the Valley of the Great Salt Lake: Including a Reconnaissance of a New Route Through the Rocky Mountains. Lippincott, Gramabo & Co, Philadelphia.

State of Utah, 2018a. https://wildlife.utah.gov/habitat/farmington_bay.php. (accessed 07.03.18).

State of Utah, 2018b. https://wildlife.utah.gov/gsl/industries/index.php. (accessed 9.03.18).

Stephens, D.W., 1974. A summary of biological investigations concerning the Great Salt Lake, Utah (1861–1973). Great Basin Natur. 34 (3). Article 7.

Stephens, D.W., 1990. Changes in lake levels, salinity and the biological community of Great Salt Lake (Utah, USA), 1847–1987. Dev. Hydrobiol. 59, 139–146. https://doi.org/10.1007/BF0002694.

Stephens, D.W., 1998. Salinity-induced changes in the aquatic ecosystems of Great Salt Lake, Utah. In: Pitman, J., Carroll, A. (Eds.), Modern and Ancient Lake Systems. In: Utah Geological Survey Guidebook, vol. 26. State of Utah Department of Natural Resources, Salt Lake City, Utah, pp. 1–7.

Stephens, D.W., Gillespie, D.M., 1972. Community structure and ecosystem analysis of the Great Salt Lake. In: Riley, J.P. (Ed.), The Great Salt Lake and Utah's Water Resources: Proceedings of the First Annual Conferernce of Utah Section American Water Resources Association. Utah Water Research Laboratory, Utah State University, Logan, UT, pp. 66–72.

Stephens, D.W., Gillespie, D.M., 1976. Phytoplankton production in the Great Salt Lake, Utah, and a laboratory study of algal response to enrichment. Limnol. Oceanogr. 21, 74–87. https://doi.org/10.4319/lo.1976.21.1.0074.

Stube, J.C., Post, F.J., Procella, D.B., 1976. Nitrogen Cycling in Microcosms and Application to the Biology of the North Arm of Great Salt Lake. (Publication No. PRJSBA-016-1). Utah Water Research Laboratory, Utah State University, Logan, UT.

Sturm, P.A., 1980. The Great Salt Lake brine system. In: Gwynn, J.W. (Ed.), Great Salt Lake: A Scientific, Historical and Economic Overview. Utah Geological Survey, State of Utah Department of Natural Resources, Salt Lake City, UT, pp. 147–162.

Stutz, J., Ackermann, R., Fast, J.D., Barrie, L., 2002. Atmospheric reactive chlorine and bromine at the Great Salt Lake, Utah. Geophys. Res. Lett. 29 (10). https://doi.org/10.1029/2002GL014812.

Tayler, P.L., Hutchinson, L.A., Muir, M.K., 1980. Heavy metals in the Great Salt Lake, Utah. Utah Geol. Mineral Survey Bull. 116, 95–200.

Tazi, L., Breakwell, D.P., Harker, A.R., Crandall, K.A., 2014. Life in extreme environments: microbial diversity in Great Salt Lake, Utah. Extremophiles 18, 525–535.

Tepšič, K., Gunde-Cimerman, N., Frisvad, J.C., 1997. Growth and mycotoxin production by *Aspergillus fumigatus* strains isolated from a saltern. FEMS Microbiol. Lett. 157, 9–12.

Tilden, J., 1898. American Algae. *Cent. III* No. 298.

Tkavc, R., 2012. Microbial Communities of the Brine Shrimp *Artemia* sp and Selected Hypersaline Microbial Mats (Doctoral dissertation). University of Ljubljana, Slovenia.

Tosca, N.J., Knoll, A.H., McLennan, S.M., 2008. Water activity and the challenge for life on early Mars. Science 320 (5880), 1204–1207.

Tsai, C.R., Garcia, J.L., Patel, B.K., Cayol, J.L., Baresi, L., Mah, R.A., 1995. *Haloanaerobium alcaliphilum* sp. nov., an anaerobic moderate halophile from the sediments of Great Salt Lake, Utah. Int. J. Syst. Bacteriol. 45, 301–307.

Turk, L.J., 1970. Evaporation of Brine: a field study on the Bonneville Salt Flats, Utah. Water Resour. Res. 6 (4), 1209–1215. https://doi.org/10.1029/WR006i004p01209.

Underwood, G., Provot, L., 2000. Determining the environmental preferences of four estuarine epipelic diatom taxa: growth across a range of salinity, nitrate and ammonium conditions. Eur. J. Phycol. 35 (2), 173–182.

United States Bureau of Reclamation, 1962. Bear River Project, Part I, Feasibility Report, Oneida Division, Idaho and Utah; Part II, Reconnaissance Report, Blacksmith Fork Division, Utah (86). United States Bureau of Reclamation, Salt Lake City, Utah.

United States Division of Fish and Wildlife, 2018. https://www.fws.gov/Refuge/Bear_River_Migratory_Bird_Refuge/about.html. (accessed 7.02.18).

United States Geologic Survey, 2018a. http://ut.water.usgs.gov/greatsaltlake/elevations/. (accessed 7.02.18).

United States Geologic Survey, 2018b. https://www.usgs.gov/media/videos/new-breach-allows-flow-great-salt-lake/. (accessed 7.02.18).

Van Auken, O.W., McNulty, L.B., 1973. The effect of environmental factors on the growth of a halophylic species of algae. Biol. Bull. 145, 210–222.

Van den Kerkhof, A.M., Hein, U.F., 2001. Fluid inclusion petrography. Lithos 55 (1), 27–47.

Vanden Berg, M.D., Chidsey, T.C., Eby, D.E., 2016. Characterization of microbialites from Antelope Island's Bridger Bay and Promontory Point, Great Salt Lake, Utah. In: Presentation at the American Association of Petroleum Geologists, Annual Convention and Exhibition, June 2016, Calgary, Alberta, Canada.

Ventosa, A., Gutierrez, M.C., Kamekura, M., Dyall-Smith, M.L., 1999. Proposal to transfer *Halococcus turkmenicus*, *Halobacterium trapanicum* JCM 9743 and strain GSL-11 to *Haloterrigena turkmenica* gen. nov., comb. nov. Int. J. Syst. Bacteriol. 49, 131–136.

Ventosa, A., de la Haba, R.R., Sánchez-Porro, C.R., Papke, R.T., 2015. Microbial diversity of hypersaline environments: a metagenomic approach. Curr. Opin. Microbiol. 25, 80–87. https://doi.org/10.1016/j.mib.2015.05.002.

Verrill, A.E., 1869. Territories of Wyoming and Idaho (1878). In: U.S. Geological & Geographical Survey Annual Report 12 Pt. 1. U.S. Government Printing Office, Washington, DC.

Vorhies, C.T., 1917. Notes on the fauna of the Great Salt Lake. Am. Nat. 61, 494–499.

Vreeland, R.H., Rosenzweig, W.D., Powers, D.W., 2000. Isolation of a 250 million-year-old halotolerant bacterium from a primary salt crystal. Nature 407, 897–900.

Wainø, M., Tindall, B.J., Schumann, P., Ingvorsen, K., 1999. *Gracilibacillus* gen nov., with description of *Gracilibacillus halotolerans* gen. nov., sp. nov.; transfer of *Bacillus dipsosauri* to *Gracilibacillus dipsosauri* comb. nov., and *Bacillus salexigens* to the genus

Salibacillus gen. nov., as *Salibacillus salexigens* comb. nov. Int. J. Syst. Bacteriol. 49, 821–831.

Ward, D.M., Brock, T.D., 1978. Hydrocarbon biodegradation in hypersaline environments. Appl. Environ. Microbiol. 35, 353–359.

Weimer, B.C., Rompato, G., Parnell, J., Gann, R., Ganesan, B., Navas, C., Gonzalez, M., Clavel, M., Albee-Scott, S., 2009. Microbial biodiversity of Great Salt Lake, Utah. Nat. Resources Environ. Issues 15, 15–22.

White, A.L., Jahnke, L.S., 2002. Contrasting effects of UV-A and UV-B on photosynthesis and photoprotection of β-carotene in two *Dunaliella* spp. Plant Cell Physiol. 43 (8), 877–884.

Wierzchos, J., Ascaso, C., McKay, C.P., 2006. Endolithic cyanobacteria in halite rocks from the hyperarid core of the Atacama Desert. Astrobiology 6 (3), 415–422. https://doi.org/10.1089/ast.2006.6.415.

Willerslev, E., Hansen, A.J., Rønn, R., Brand, T.B., Barnes, I., Wiuf, C., Gilichinsky, D., Mitchell, D., Cooper, A., 2004. Long-term persistence of bacterial DNA. Curr. Biol. 14 (1), R9–R10.

Winters, Y.D., 2013. Haloarchaeal Survival and Preservation of Biomaterials (Carotenoids) in Ancient Halite (Doctoral dissertation). State University of New York at Binghamton, Binghamtib, NY.

Woese, C.R., Fox, G.E., 1977. Phylogenetic structure of the prokaryotic domain: the primary kingdoms. Proc. Natl. Acad. Sci. U. S. A. 74 (11), 5088–5090.

Wurtsbaugh, W.A., 1988. Iron, molybdenum and phosphorus limitation of N2 fixation maintains nitrogen deficiency of plankton in the Great Salt Lake drainage (Utah, USA). Verhandlungen. Int. Vereinigung. Theor. Angew. Limnol. 23, 121–130.

Wurtsbaugh, W.A., 1992. Food-web modification by an invertebrate predator in the Great Salt Lake (USA). Oecologia 89 (2), 168–175.

Wurtsbaugh, W., 2007. Preliminary analyses of selenium bioaccumulation in benthic food webs of the Great Salt Lake, Utah. In: Final Report: Development of a Selenium Standard for the Open Waters of the Great Salt Lake. Utah Department of Environmental Quality, Salt Lake City, UT.

Wurtsbaugh, W.A., 2009. Biostromes, brine flies, birds and the bioaccumulation of selenium in Great Salt Lake, Utah. Natl. Resour. Environ. 15, 2.

Wurtsbaugh, W.A., Berry, T.S., 1990. Cascading effects of decreased salinity on the plankton, chemistry, and physics of the Great Salt Lake (Utah). Can. J. Fish. Aquat. Sci. 47, 100–109.

Wurtsbaugh, W.A., Gliwicz, Z.M., 2001. Limnological control of brine shrimp population dynamics and cyst production in the Great Salt Lake, Utah. Hydrobiologia 466, 119–132.

Wurtsbaugh, W., Naftz, D.L., Bradt, S., 2009. Eutrophication, nutrient fluxes and connectivity between the bays of Great Salt Lake, Utah (abs.). In: Oren, A., Naftz, D., Palacios, P., Wurtsbaugh, W.A. (Eds.), Saline Lakes Around the World: Unique Systems with Unique Values, Natural Resources and Environmental Issues. In: vol. 15. S.J. and Jessie E. Quinney Natural Resources Research Library, Logan, Utah, p. 51.

Wurtsbaugh, W.A., Gardberg, J., Izdepski, C., 2011. Biostrome communities and mercury and selenium bioaccumulation in the Great Salt Lake (Utah, USA). Sci. Total Environ. 409 (20), 4425–4434.

Wurtsbaugh, W.A., Miller, C., Null, S.E., DeRose, R.J., Wilcock, P., Hahnenberger, M., Howe, F., Moore, J., 2017. Decline of the world's saline lakes. Nat. Geosci. 10 (11), 816.

Zajc, J., Zalar, P., Gunde-Cimerman, N., 2017. Yeasts in hypersaline habitats. In: Buzzini, P., Lachance, A., Yurkov, A. (Eds.), Yeasts in Natural Ecosystems: Ecology and Diversity. Springer, Dordrecht, Netherlands.

Zalar, P., Kocuvan, M.A., Plemenitaš, A., Gunde-Cimerman, N., 2005. Halophilic black yeasts colonize wood immersed in hypersaline water. Bot. Mar. 48, 323–326.

Zalar, P., De Hoog, G.S., Schroers, H.-J., Crous, P.W., Groenewald, J.Z., Gunde-Cimerman, N., 2007. Phylogeny and ecology of the ubiquitous saprobe *Cladosporium sphaerospermum*, with descriptions of seven new species from hypersaline environments. Stud. Mycol. 58, 157–183. https://doi.org/10.3114/sim.2007.58.06.

Zeikus, J.G., Hegge, P.W., Thompson, T.E., Phelps, T.J., Langworthy, T.A., 1983. Isolation and description of *Haloanaerobium praevalens* gen. nov. sp. nov., and obligately anaerobic halophile common to Great Salt Lake sediments. Curr. Microbiol. 9, 225–234.

Zobell, C.E., 1937. Direct microscopic evidence of an Autochthonous bacterial flora in Great Salt Lake. Ecology 18 (3), 453–458.

Zvyagintseva, I.S., Tarasov, A.L., 1987. Extreme halophilic bacteria from saline soils. Microbiology (English translation of Mikrobiologiia) 56, 664–669.

Plants and salt: Plant response and adaptations to salinity

Nirit Bernstein

Institute of Soil, Water and Environmental Sciences, ARO, Volcani Center, Rishon LeZion, Israel

Chapter outline

1 Introduction

Salt-affected soils are abundant in semiarid and arid regions of the world. Saliniza-tion of soils occurs through natural processes including evaporation of saline under-ground water, sea water infiltration of coastal ground waters, sea water salts in wind and rain, as well as human-induced processes, such as irrigation with marginal water and poor agrotechniques. This results in the accumulation of dissolved salts in the soil and water to an extent that restricts plant growth and agricultural crop produc-tion. More than 800 million ha are salt affected, which is more than 12% of the world land area, and more than 20% of cultivated land is salt affected (FAO, 1994; Munns and Tester, 2008). A soil is considered saline when the electric conductivity, EC, of the saturated soil solution reaches 4 dS m^{-1}, which is equivalent to about 40 mM NaCl. This level of salinity generates an osmotic pressure of about 0.2 MPa and reduces growth and yields of most crop plants (Munns and Tester, 2008).

A major contributor to salinity damages to plants are irrigation water which con-tain above optimal concentrations of solutes, specifically Na$^+$ and Cl$^-$. Due to short-age of fresh waters in large parts of the world, marginal water of high salinity, mainly

Model Ecosystems in Extreme Environments. https://doi.org/10.1016/B978-0-12-812742-1.00005-2

saline water or recycled treated effluents are increasingly used for irrigation (Bernstein, 2009). These water often contain higher levels of Na and Cl than optimal for plant growth and development and their utilization for irrigation requires adjustment of agrotechniques to maintain quality production. Water is essential for plant survival. They are vital for all life required metabolic process in the cells. Additionally, bulk flow of water in the plant facilitates long- and short-distance translocation, such as of carbohydrates from the photosynthetic production sites at the shoot throughout to the roots, and transport of minerals from the uptake sites at the roots to the shoot. Although there is an abundance of water available on the planet, most of it is not of quality suitable to sustain plant life. Most of the Earth surface, about 72%, is covered in seawater, which is a salt solution dominated by Na^+ and Cl^-. Sea and ocean waters usually contain about 560 mM Cl and 480 mM Na. These water also contain substantial concentrations of other ions which are essential for plant growth: Mg^{2+} (55 mM), SO_4^{2-} (29 mM), K (10 mM), and Ca^{2+} (10 mM), and very low concentrations of PO_4^-/PO_4^{2-} and NO_3 (Harvey, 1966). Although the concentrations of most ions in the oceans are not in the optimal range for plants, what makes them unsuitable for growth of most plants is the high concentrations of Na and Cl.

Across the plant kingdom exists a wide range of tolerance to Na and Cl salinity. Depending on the sensitivity of plants to salinity, and the ability to grow in saline environments, they are classified as either glycophytes (that are characterized by low tolerance to salinity) or halophytes (that thrive under high salinity level). Most crop species belong to the first category. The tolerance level to salinity ranges from very sensitive, for example, chickpea, to the very tolerant halophytes (e.g., euhalophytes) (Breckle, 2002; Flowers et al., 2010; English and Colmer, 2013), vascular plants that can thrive under long-term exposure to seawater in the root zone (Flowers and Colmer, 2008).

Excessive accumulation of salts in the root growing solution and irrigation with saline water alters water relations in the plant and can lead to excessive, damaging, accumulation of salts *in planta*. As a consequence, ion toxicity often induces necrosis or chlorosis in the plant tissues, frequently due to Na^+ but also Cl^- accumulation that interfere with a range of physiological processes in the plant. The damaging effects of salinity depend on the plant species and environmental factors such as temperature, humidity, light intensity, and soil conditions (Bernstein, 2013; Tang et al., 2015).

Salinity has a different impact on growth and development of different plant organs, leading to altered plant morphology (Bernstein, 2013; Bernstein et al., 2010; Julkowska and Testerink, 2015). Plants have developed different physiological and biochemical mechanisms in order to cope with the resulting stress condition. Such adaptive mechanisms include changes in water relations, anatomy, morphology, biochemical adaptation, compartmentation or exclusion of toxic ion (Parida and Das, 2005; Ashraf and Harris, 2013; Bernstein, 2013; Acosta-Motos et al., 2017a, 2017b).

Our knowledge of survival mechanisms under extreme environments including saline habitats and the limits of life on Earth has greatly advanced. This progress has given us important directions concerning the origins of life on Earth and has been instrumental to the development of astrobiology that studies the evolution and distribution of life in the universe.

2 Mechanisms of salinity response

Salt stress effects on plants can be divided into first-occurring osmotic stress and later occurring toxicity/specific ion stress due to accumulation of ionic Na and Cl (Munns and Tester, 2008). Accumulation of salts in the soil and irrigation with saline water decrease the water potential of the soil solution and restrict water uptake by the plant. High concentrations of Na in the soil solution affect uptake of essential mineral nutrients such as K and Ca by the roots and their availability to plant cells (Lazof and Bernstein, 1998, 1999). Once concentrations of Na and Cl in the plant reach a damaging threshold level, specific toxicities as well as alterations in water relations are further induced (Munns and Tester, 2008). Other effects of salinity include impaired metabolism and oxidative damage (Bernstein et al., 2010). The combined response is restriction of plant growth and development and reduced productivity (Bernstein, 2013; Bernstein et al., 2010).

Plants evolved a variety of physiological mechanisms to enable survival under salinity conditions. The main mechanisms involve osmotic adjustment to mitigate osmotic stress; ion exclusion, inclusion, and compartmentation for minimizing the damaging ion-specific effects of Na and nutrient deficiencies; antioxidant response; hormonal regulation. These are discussed as follows.

3 Water relations and osmotic adjustment: The role of osmolytes

Under exposure to salinity, the plant adjusts its water potential by losing water, which causes a decrease in osmotic potential and turgor. This induces a signal that triggers adaptive responses (Hasegawa et al., 2000). During the adjustment period, the cells accumulate organic and inorganic solutes that reduce the osmotic potential and the hydraulic conductivity of the membranes is reduced. Tissue growth proceeds once turgor is recovered (Kaldenhoff et al., 2008; Munns and Tester, 2008).

Common responses of plants to salinity exposure are an increase in solute concentration, for example, osmotic adjustment, changes to the cell wall elasticity, decrease in relative water content in the tissue, and increase in the percentage of water in the apoplast, which reduces salinity damages by maintaining turgidity of the tissue. Accumulation of organic osmolytes in plant cells is key to maintaining low intracellular osmotic potential of plants and thereby preventing the damaging effects of salinity (Verslues et al., 2006). Common osmolytes in vascular plants include proline, glycine-betaine, polyamines, sugar alcohols, and proteins from the LEA superfamily. Information about the regulatory networks of osmolyte biosynthesis and accumulation is important for understanding plant salinity tolerance. Salinity stress stimulates the synthesis of osmolytes. The extent of osmotic adjustment differs between plant species and may be affected by a variety of factors including plant age and rate of stress application (Stepien and Johnson, 2009).

Among the organic osmolytes, proline and glycine-betaine are the most abundant and efficient compatible solutes (Tang et al., 2015). Proline synthesis is enhanced under salinity and its catabolism is enhanced during recovery from the stress (Sharma and Verslues, 2010), and prevention of proline synthesis increases sensitivity to salinity (Szekely et al., 2008). In addition to its conventionally assumed role in osmotic adjustment, proline is now considered to act as a reactive oxygen scavenger, redox buffer, and molecular chaperone and stabilizes membranes and proteins under stress conditions (Verbruggen and Hermans, 2008). Glycine-betaine as well in addition to its function as compatible solute is considered to involve in protection of enzymes and membrane structures (Guinn et al., 2011).

4 Ion exclusion, inclusion, and compartmentation

Salt stress damages are often associated with toxic effects caused by the uptake of salt ions mainly Na and Cl by the root and their accumulation in the plant. Some mechanisms of salt tolerance to salinity include strategies for exclusion of ions from salt-sensitive organs or subcellular compartments/tissues by their inclusion in less sensitive locals such as the root, old leaves, or vacuoles.

High concentrations of ions in the cytoplasm are damaging to plants. Excess ionic minerals are transported across the tonoplast (the membrane surrounding the vacuole) and compartmentalized in vacuoles, thereby protecting the plant from salinity stress (Zhu, 2003). The Na is transported from the cytoplasm to the vacuole by a Na/H antiporter. Two types of H^+ pumps are found in the tonoplast, V-H+-ATPase (V-ATPase) and the pyrophosphatase (V-PPase), that energize secondary transport and are therefore important for maintaining solute homeostasis. Under salinity their activity may increase to a different extent in different plants (Dietz et al., 2001; Wang et al., 2001).

Other exclusion strategies involve accumulation of salts in the root and hence their exclusion from the aerial part of the plant. This mechanism occurs in most of the halophytes and in some glycophytes. Saline vesicle glands in the epidermis of some organs, usually leaves or stems, facilitate excretion of salts from the cell, preventing salt accumulation. Another exclusion mechanism involves sequestering salts in older tissues such as old leaves, which causes salt-induced senescence (Reddy et al., 1992).

In some plants, especially halophytes, compartmentation of Na is part of an inclusion strategy. Na includers may accumulate salts to be served as osmoticum to lower the osmotic potential to drive up water uptake for the maintenance of leaf turgor

(Koyro, 2006). The energetic cost of intracellular ion compartmentation is considered low compared with the synthesis of organic osmolytes (Munns, 2002).

Numerous plant membrane transporters play a key role in mechanisms of resistance to salt stress, particularly Na and K transporters (Schroeder et al., 2013). Influx of Na into roots across the plasma membrane can occur via ion channels and transporters. Various Ca-permeable channels are permeable to Na as well and can facilitate entry of Na into the cell (NSCCs: Tyerman and Skerrett, 1999; Tester and Davenport, 2003; CNGC: Hua et al., 2003; GLR: Tapken and Hollmann, 2008). Na can enter cells via Na transporters (Horie et al., 2007), and a cation/H antiporter in the root endodermis is involved in Na transport from the endodermis to the stele (Hall et al., 2006). Potential candidates for xylem loading of Na are the outward-rectifying K channels KORC and NORC (de Boer and Wegner, 1997; Wegner and de Boer, 1997), and class I HKT transporters remove Na from the xylem (Sunarpi et al. (2005).

Concentrations of K and Ca in the plant are often reduced under salinity due to competition for uptake with Na (Lazof and Bernstein, 1998, 1999) and consequently, salt-stressed plants often suffer from Ca deficiency (Neves-Piestun and Bernstein, 2005). Optimal function under salinity requires adequate K supply to the cells. Inward-rectifying K channels and outward-rectifying K channels that function in K-selective influx and efflux in plant cells, respectively, may also reduce Na toxicity (Deinlein et al., 2014). Maintenance of K uptake with exclusion of Na from leaves correlates with plant salt tolerance (Blumwald and Poole, 1985).

The Na/K transporters, a product of the HKT genes, have a major role in plant salt tolerance and root/shoot partitioning of Na (Horie et al., 2009). HKT class I transporters mediate more Na-selective transport (Maser et al., 2002) and class II, Na-K cotransport (Rubio et al., 1995). A tonoplast Na/H exchanger (NHX1) (Blumwald and Poole, 1985) and plasma membrane Na/H antiporter (SOS1 = NHX7) (Yamaguchi et al., 2013) maintain low Na concentrations of Na in the cytoplasm, by compartmentalizing Na in the vacuole, and export of Na out of the cell. Ion transport regulation mechanism under salinity was further reviewed by Deinlein et al. (2014).

5 Antioxidant defense response

Salinity induces oxidative stress in plants (Hernández et al., 2001; Mittova et al., 2004; Bernstein et al., 2010). Increased accumulation of reactive oxygen species (ROS) such as superoxide radicals and hydrogen peroxide occurs in different cell compartments under salinity in correlation with an increase in oxidative stress factors, such as lipid peroxidation and protein oxidation (Mittova et al., 2004; Bernstein et al., 2010; Acosta-Motos et al., 2017a, 2017b). The extent of salinity tolerance was reported for many, but not all plants to correlate with the extent of the response of the antioxidative defense system. Up-regulation of the antioxidative system in response to salinity exposure (Bernstein et al., 2010; Rubio et al., 2009), higher constitutive levels of some antioxidant enzymes (Tsugane et al., 1999), and increase in the antioxidant defenses on the subcellular level such as the chloroplasts and mitochondria (Gómez et al., 1999) are some of the mechanisms involved in the salt tolerance response. Salt-sensitive species show an unchanged/decreased response or lower constitutive antioxidant enzyme levels than salt-tolerant species (Mittova et al., 2003, 2004). Differences between the oxidative response of roots and leaf cells to salinity are responsible for the higher sensitivity to salinity of leaves compared to roots (Bernstein et al., 2010). Differential involvement of scavenging enzymes in cell growth restriction under salinity occurs throughout cell development (Bernstein et al., 2010; Kravchik and Bernstein, 2013). The antioxidative response is involved in the ameliorative effects of Ca on cell growth restriction under salinity (Shoresh et al., 2011) through modulating activity of ROS producing enzymes and ROS levels. Enhancement of polyamine oxidase activity in conjunction with low levels of apoplastic peroxidase by supplemental Ca suggests cellular growth maintenance via nonenzymatic wall loosening, derived by the increase in H_2O_2, rather than by peroxidase-mediated cross-linking of wall material. Thus extracellular Ca can modulate ROS level in specific locales and developmental stages thereby affecting cellular extension (Shoresh et al., 2011). Effect of salinity on apoplast acidification kinetics in the growing zone disclaimed salt stress effects on "acid growth" (Neves-Piestun and Bernstein, 2001).

Nitric oxide (NO) is also involved in the salinity response of plants. Pretreatment with NO donor, or exogenous application of NO, can improve plants' response to salinity (Dinler et al., 2014; Kaur and Bhatla, 2016). NO can also act as a regulator of glutathione metabolism, modulating the activity of GSH-dependent enzymes such as GST, GR, and GSH contents (Dinler et al., 2014; Kaur and Bhatla, 2016; Manai et al., 2014). It was suggested that under salinity stress, plant peroxisomes, in addition to generating ROS, are a source of NO production (Corpas et al., 2009).

6 Hormonal regulation and salt tolerance

Exposure to salinity triggers an increase in levels of plant hormones such as ABA and cytokinins which are considered to involve in the tolerance response to salinity (Thomas et al., 1992; Aldesuquy, 1998; Vaidyanathan et al., 1999). ABA is upregulated in plants under conditions of water deficit and is well known to involve in response to water stress. Salinity stress causes osmotic stress which induces production of ABA in roots and shoots. The increased accumulation of the hormone, and exogenous application, can reduce negative effects of salinity on growth, photosynthesis, and translocation of assimilates (He and Cramer, 1996; Cabot et al., 2009; Cramer and Quarrie, 2002; Kang et al., 2005; Popova et al., 1995).

ABA involvement in the salinity response is related also to the accumulation of compatible solutes and nutritional cations K^+ and Ca^{2+} in vacuoles of roots (Chen et al., 2001; Gurmani et al., 2012). It modulates the expression of a number of salinity responsive genes, including *HVP1* and *HVP10*, for vacuolar H^+-inorganic pyrophosphatase, and of *HvVHA-A*, for subunit A of vacuolar H^+-ATPase in *Hordeum vulgare* (Fukuda et al., 2004).

Jasmonates, salicylic acid, and brassinosteroids are also involved in the response to salinity (Fragnire et al., 2011; Clause and Sasse, 1998; Jayakannan et al., 2013). Current knowledge and the potential of exogenous application to reduce damaging effects of salinity in plants were reviewed by Ashraf et al. (2010).

7 Sensing salt-stress

The ability to response to a stress condition requires the ability to sense the imposed stress. High concentration of salts in the root growing zone imposes a hyperosmotic stress in the roots. Plants have evolved mechanisms to sense both components of the salinity stress, that is, the ion-specific component (high Na^+ concentration), as well as the hyper-osmotic component. This two-component mechanism of salinity stress sensing is evident by the existence of some distinct responses to NaCl, which differ from responses to osmotic stress per se. The molecular identities of plant hyperosmotic sensors and Na^+ sensors are yet unclear (reviewed by Deinlein et al., 2014). Some evidence suggests a role of osmosensing to HK1 (Urao et al., 1999; Tran et al., 2007; Wohlbach et al., 2008). But, some of the physiological responses to osmotic stress remain unaffected in an hk1 mutant, suggesting that other proteins must still be sensing the osmotic stress (Kumar et al., 2013). Hyper-osmotic stress was hypothesized to be sensed by a mechanically gated Ca^{2+} channel (Kurusu et al. (2013) because a rapid rise in cytosolic Ca^{2+} occurs following exposure to NaCl or mannitol (Knight et al., 1997). Mutations that affect cuticle development were indeed demonstrated to interfere with numerous osmotic-induced responses, including production of ABA (Wang et al., 2011). Deinlein et al. (2014) pointed out that the cuticle provides structural support to the plasma membrane and the cell wall and could alter the diffusibility of water into the cell, therefore, changing cuticle properties may affect mechanical properties of water stress on the cell. Other second messengers that are linked to Ca^{2+} signaling are also induced by hyperosmotic or salt stress, for example, reactive oxygen species (Jiang et al., 2013). Although a rapid increase in Ca^{2+} is a well-known response to osmotic stress, Ca^{2+}-independent osmotic sensory mechanisms were also suggested to exist (Deinlein et al., 2014).

8 Summary

Plants are adaptable organisms. They have the potential for adaptation to extreme environments. This has facilitated plant colonization of most niches on earth. However, most plants cannot survive under conditions of high salinity. Natural processes and manmade activity continuously increase the salinization of soils, making salinity a major abiotic factor threatening food security throughout the world. Mechanisms of salinity tolerance encompass a range of responses at the morphological, physiological, molecular, and metabolic levels, and salinity exposure induces signaling pathways directed to induce acclimation. Despite the large increase in our understanding of the salinity response and tolerance mechanisms at the physiological, molecular, metabolic, and genomic levels, salinity continues to decrease crop production, emphasizing the need for a combined approach for the identification of key components in pathways controlling salinity tolerance to be used for development of tolerant crops.

In the last decades, a substantial increase in our understanding of plants' and other organisms' tolerance to extreme conditions facilitated a broader understanding of what we now consider limits to habitable environmental conditions. Our knowledge of mechanisms involved in tolerance of extremophiles to extreme environments has been greatly expanded. This progress was fundamental for the development of studies into the origin, evolution, and spread of life in extreme niches in the universe.

References

Acosta-Motos, J.R., Hernández, J.A., Álvarez, S., Barba-Espín, G., Sánchez-Blanco, M.J., 2017a. Long-term resistance mechanisms and irrigation critical threshold showed by *Eugenia myrtifolia* plants in response to saline reclaimed water and relief capacity. Plant Physiol. Biochem. 111, 244–256.

Acosta-Motos, J.R., Ortuño, M.F., Bernal-Vicente, A., Diaz-Vivancos, P., Sanchez-Blanco, M.J., Hernandez, J.A., 2017b. Plant responses to salt stress: adaptive mechanisms. Agronomy 7, 18. https://doi.org/10.3390/agronomy7010018.

Aldesuquy, H.S., 1998. Effect of seawater salinity and gibberllic acid on abscisic acid, amino acids and water-use efficiency of wheat plants. Agrochimica 42, 147–157.

Ashraf, M., Harris, P.J.C., 2013. Photosynthesis under stressful environments: an overview. Photosynthetica 51, 163–190.

Ashraf, M., Akram, N.A., Arteca, R.N., Foolad, M.R., 2010. The physiological, biochemical and molecular roles of brassinosteroids and salicylic acid in plant processes and salt tolerance. Crit. Rev. Plant Sci. 29, 162–190.

Bernstein, N., 2009. Contamination of soils with microbial pathogens originating from effluent water used for irrigation. In: Contaminated Soils: Environmental Impact, Disposal and Treatment. Nova Science Publishers, NY, USA, pp. 473–486.

Bernstein, N., 2013. Effects of salinity on root growth. In: Eshel, A., Beeckman, T. (Eds.), Plant Roots: The Hidden Half. fourth ed. CRC Press, Boca Raton. 848 p.

Bernstein, N., Shoresh, M., Xu, Y., Huang, B., 2010. Involvement of the plant antioxidative response in the differential growth sensitivity to salinity of leaves vs. roots during cell development. Free Rad. Biol. Med. 49, 1161–1171.

Blumwald, E., Poole, R.J., 1985. Na/H antiport in isolated tonoplast vesicles from storage tissue of Beta vulgaris. Plant Physiol. 78, 163–167.

Breckle, S.W., 2002. Salinity, halophytes and salt affected natural ecosystems. In: Läuchli, A., Lüttge, U. (Eds.), Salinity: Environment-Plants-Molecules. Springer, Dordrecht, pp. 53–77.

Cabot, C., Sibole, J.V., Barcel'o, J., Poschenrieder, C., 2009. Abscisic acid decreases leaf Na+ exclusion in salt-treated *Phaseolus vulgaris* L. J. Plant Growth Regul. 8, 187–192.

Chen, S., Li, J., Wang, S., Hüttermann, A., Altman, A., 2001. Salt, nutrient uptake and transport, and ABA of *Populus euphratica*; a hybrid in response to increasing soil NaCl. Trees—Struct. Funct. 15, 186–194.

Clause, S.D., Sasse, J.M., 1998. Brassinosteroids: essential regulators of plant growth and development. Annu. Rev. Plant Biol. 49, 427–451.

Corpas, F.J., Hayashi, M., Mano, S., Nishimura, M., Barroso, J.B., 2009. Peroxisomes are required for in vivo nitric oxide accumulation in the cytosol following salinity stress of Arabidopsis plants. Plant Physiol. 151, 2083–2094.

Cramer, G.R., Quarrie, S.A., 2002. Abscisic acid is correlated with the leaf growth inhibition of four genotypes of maize differing in their response to salinity. Funct. Plant Biol. 29, 111–115.

de Boer, A.H., Wegner, L.H., 1997. Regulatory mechanisms of ion channels in xylem parenchyma cells. J. Exp. Bot. 48, 441–449.

Deinlein, U., Stephan, A.B., Horie, T., Luo, W., Xu, G., Schroeder, J.I., 2014. Plant salt-tolerance mechanisms. Trends Plant Sci. 19, 371–379.

Dietz, K.J., Tavakoli, N., Kluge, C., et al., 2001. Significance of the Vtype ATPase for the adaptation to stressful growth conditions and its regulation on the molecular and biochemical level. J. Exp. Bot. 52, 1969–1980.

Dinler, B.S., Antoniou, C., Fotopoulos, V., 2014. Interplay between GST and nitric oxide in the early response of soybean (*Glycine max* L.) plants to salinity stress. J. Plant Physiol. 171, 1740–1747.

English, J.P., Colmer, T.D., 2013. Tolerance of extreme salinity in two stem-succulent halophytes (*Tecticornia* species). Funct. Plant Biol. 40, 897–912.

FAO, 1994. Land and Plant Nutrition Management Service.

Flowers, T.J., Colmer, T.D., 2008. Salinity tolerance in halophytes. New Phytol. 179, 945–963.

Flowers, T.J., Gaur, P.M., Gowda, C.L.L., et al., 2010. Salt sensitivity in chickpea. Plant Cell Environ. 33, 490–509.

Fragnire, C., Serrano, M., Abou-Mansour, E., M'etraux, J.-P., L'Haridon, F., 2011. Salicylic acid and its location in response to biotic and abiotic stress. FEBS Lett. 58512, 1847–1852.

Fukuda, A., Chiba, K., Maeda, M., Nakamura, A., Maeshima, M., Tanaka, Y., 2004. Effect of salt and osmotic stresses on the expression of genes for the vacuolar H^+-pyrophosphatase, H^+ ATPase subunit A, and Na^+/H^+ antiporter from barley. J. Exp. Bot. 55, 585–594.

Gómez, J.M., Hernández, J.A., Jiménez, A., del Río, L.A., Sevilla, F., 1999. Differential response of antioxidative enzymes of chloroplasts and mitochondria to long-term NaCl stress of pea plants. Free Rad. Res. 31, S11–S18.

Guinn, E.J., Pegram, L.M., Capp, M.W., Pollock, M.N., Record, M.T.E., 2011. Quantifying why urea is a protein denaturant, whereas glycine betaine is a protein stabilizer. Proc. Natl. Acad. Sci. U. S. A. 108, 16932–16937.

Gurmani, A.R., Bano, A., Khan, S.U., Din, J., Zhang, J.L., 2012. Alleviation of salt stress by seed treatment with abscisic acid (ABA), 6-benzylaminopurine (BA) and chlormequat chloride. Biochim. Biophys. Acta—Gene Regul. Mech. 1819, 86–96.

Hall, D., Evans, A.R., Newbury, H.J., Pritchard, J., 2006. Functional analysis of CHX21: a putative sodium transporter in Arabidopsis. J. Exp. Bot. 57, 1201–1210.

Harvey, H.W., 1966. The Chemistry and Fertility of Sea Waters. Cambridge University Press, Cambridge.

Hasegawa, P.M., Bressan, R.A., Zhu, J.K., Bhnert, H.J., 2000. Plant cellular and molecular responses to high salinity. Ann. Rev. Plant Physiol. Plant Mol. Biol. 51, 463–499.

He, T., Cramer, G.R., 1996. Abscisic acid concentrations are correlated with leaf area reductions in two salt-stressed rapid cycling *Brassica* species. Plant Soil 179, 25–33.

Hernández, J.A., Ferrer, M.A., Jiménez, A., Ros-Barceló, A., Sevilla, F., 2001. Antioxidant systems and O_2/H_2O_2 production in the apoplast of *Pisum sativum* L. leaves: its relation with NaCl-induced necrotic lesions in minor veins. Plant Physiol. 127, 817–831.

Horie, T., Costa, A., Kim, T.H., Han, M.J., Horie, R., Leung, H.Y., Miyao, A., Hirochika, H., An, G., Schroeder, J.I., 2007. Rice OsHKT2;1 transporter mediates large Na^+ influx component into K^+-starved roots for growth. EMBO J. 26, 3003–3014.

Horie, T., Hauser, F., Schroeder, J.I., 2009. HKT transporter-mediated salinity resistance mechanisms in Arabidopsis and monocot crop plants. Trends Plant Sci. 14, 660–668.

Hua, B.G., Mercier, R.W., Leng, Q., Berkowitz, G.A., 2003. Plants do it differently. A new basis for potassium/sodium selectivity in the pore of an ion channel. Plant Physiol. 132, 1353–1361.

Jayakannan, M., Bose, J., Babourina, O., et al., 2013. Salicylic acid improves salinity tolerance in *Arabidopsis* by restoring membrane potential and preventing salt-induced K^+ loss via aGORK channel. J. Exp. Bot. 64, 2255–2268.

Jiang, Z., Zhu, S., Ye, R., Xue, Y., Chen, A., An, L., Pei, Z.M., 2013. Relationship between NaCl and H_2O_2-induced cytosolic Ca^{2+} increases in response to stress in Arabidopsis. PLoS ONE 8, e76130.

Julkowska, M.M., Testerink, C., 2015. Tuning plant signaling and growth to survive salt. Trends Plant Sci. 20, 586–594.

Kaldenhoff, R., Ribas-Carbo, M., Flexas, J., Lovisolo, C., Heckwolf, M., Uehlein, N., 2008. Aquaporins and plant water balance. Plant Cell Environ. 31, 658–666.

Kang, D.-J., Seo, Y.-J., Lee, J.-D., et al., 2005. Jasmonic acid differentially affects growth, ion uptake and abscisic acid concentration in salt-tolerant and salt-sensitive rice cultivars. J. Agron. Crop Sci. 191, 273–282.

Kaur, H., Bhatla, S.C., 2016. Melatonin and nitric oxide modulate glutathione content and glutathione reductase activity in sunflower seedling cotyledons accompanying salt stress. Nitric Oxide 59, 42–53.

Knight, H., Trewavas, A.J., Knight, M.R., 1997. Calcium signalling in Arabidopsis thaliana responding to drought and salinity. Plant J. 12, 1067–1078.

Koyro, H.W., 2006. Effect of salinity of growth, photosynthesis, water relations and solute composition of the potential cash crop halophyte *Plantago coronopus* (L.). Environ. Exp. Bot. 56, 136–146.

Kravchik, M., Bernstein, N., 2013. Effects of salinity on the transcriptome of growing maize leaf cells points at differential involvement of the antioxidative response in cell growth restriction. BMC Genomics 16, 14–24.

Kumar, M.N., Jane, W.N., Verslues, P.E., 2013. Role of the putative osmosensor Arabidopsis histidine kinase1 in dehydration avoidance and low-water-potential response. Plant Physiol. 161, 942–953.

Kurusu, T., Kuchitsu, K., Nakano, M., Nakayama, Y., Iida, H., 2013. Plant mechanosensing and Ca^{2+} transport. Trends Plant Sci. 18, 227–233.

Lazof, D.B., Bernstein, N., 1998. The NaCl-induced inhibition of shoot growth: the case for disturbed nutrition with special consideration of calcium nutrition. Adv. Bot. Res. 29, 113–189.

Lazof, D.B., Bernstein, N., 1999. Effect of salinization on nutrient transport to lettuce leaves: consideration of leaf developmental stage. New Phytol. 144, 85–94.

Manai, J., Gouia, H., Corpas, F.J., 2014. Redox and nitric oxide homeostasis are affected in tomato (*Solanum lycopersicum*) roots under salinity-induced oxidative stress. J. Plant Physiol. 171, 1028–1035.

Maser, P., Hosoo, Y., Goshima, S., Horie, T., Eckelman, B., Yamada, K., Yoshida, K., et al., 2002. Glycine residues in potassium channel-like selectivity filters determine potassium selectivity in four-loop-per-subunit HKT transporters from plants. Proc. Natl. Acad. Sci. U. S. A. 99, 6428–6433.

Mittova, V., Tal, M., Volokita, M., Guy, M., 2003. Up-regulation of the leaf mitochondrial and peroxisomal antioxidative systems in response to salt-induced oxidative stress in the wild salt-tolerant tomato species *Lycopersicon pennellii*. Plant Cell Environ. 26, 845–856.

Mittova, V., Guy, M., Tal, M., Volokita, M., 2004. Salinity up-regulates the antioxidative system in root mitochondria and peroxisomes of the wild salt-tolerant tomato species *Lycopersicon pennellii*. J. Exp. Bot. 399, 1105–1113.

Munns, R., 2002. Comparative physiology of salt and water stress. Plant Cell Environ. 25, 239–250.

Munns, R., Tester, M., 2008. Mechanisms of salinity tolerance. Ann. Rev. Plant Biol. 59, 651–681.

Neves-Piestun, B.G., Bernstein, N., 2001. Salinity-induced inhibition of leaf elongation is not mediated by changes in cell-wall acidification capacity. Plant Physiol. 125, 1419–1428.

Neves-Piestun, B.G., Bernstein, N., 2005. Salinity induced changes in the nutritional status of expanding cells may impact leaf growth inhibition in maize. Funct. Plant Biol. 32, 141–152.

Parida, A.K., Das, A.B., 2005. Salt tolerance and salinity effects on plants: a review. Ecotoxicol. Environ. Saf. 60, 324–349.

Popova, L.P., Stoinova, Z.G., Maslenkova, L.T., 1995. Involvement of abscisic acid in photosynthetic process in *Hordeum vulgare* L. during salinity stress. J. Plant Growth Regul. 14, 211–218.

Reddy, M.P., Sanish, S., Iyengar, E.R.R., 1992. Photosynthetic studies and compartmentation of ions in different tissues of *Salicornia brachiata* Roxb. under saline conditions. Photosynthetica 26, 173–179, 1992.

Rubio, F., Gassmann, W., Schroeder, J.I., 1995. Sodium-driven potassium uptake by the plant potassium transporter HKT1 and mutations conferring salt tolerance. Science 270, 1660–1663.

Rubio, M.C., Bustos-Sanmaded, P., Clemente, R.M., Becana, M., 2009. Effects of salt stress on the expression of antioxidant genes and proteins in the model legume *Lotus japonicus*. New Phytol. 181, 851–859.

Schroeder, J.I., Delhaize, E., Frommer, W.B., Guerinot, M.L., Harrison, M.J., Herrera-Estrella, L., Horie, T., et al., 2013. Using membrane transporters to improve crops for sustainable food production. Nature 497, 60–66.

Sharma, S., Verslues, P.E., 2010. Mechanisms independent of abscisic acid (ABA) or proline feedback have a predominant role in transcriptional regulation of proline metabolism during low water potential and stress recovery. Plant Cell Environ. 33, 1838–1851.

Shoresh, M., Spivak, M., Bernstein, N., 2011. Involvement of calcium-mediated effects on ROS metabolism in regulation of growth improvement under salinity. Free Rad. Biol. Med. 51, 1221–1234.

Stepien, P., Johnson, G.N., 2009. Contrasting responses of photosynthesis to salt stress in the glycophyte Arabidopsis and the halophyte *Thellungiella*: role of the plastid terminal oxidase as an alternative electron sink. Plant Physiol. 149, 1154–1165.

Sunarpi, Horie, T., Motoda, J., Kubo, M., Yang, H., Yoda, K., Horie, R., et al., 2005. Enhanced salt tolerance mediated by AtHKT1 transporter-induced Na unloading from xylem vessels to xylem parenchyma cells. Plant J. 44, 928–938.

Szekely, G., et al., 2008. Duplicated P5CS genes of Arabidopsis play distinct roles in stress regulation and developmental control of proline biosynthesis. Plant J. 53, 11–28.

Tang, X., Mu, X., Shao, H., Wang, H., Brestic, M., 2015. Global plant-responding mechanisms to salt stress: physiological and molecular levels and implications in biotechnology. Crit. Rev. Biotechnol. 35, 425–437.

Tapken, D., Hollmann, M., 2008. Arabidopsis thaliana glutamate receptor ion channel function demonstrated by ion pore transplantation. J. Mol. Biol. 383, 36–48.

Tester, M., Davenport, R., 2003. Na^+ tolerance and Na^+ transport in higher plants. Ann. Bot. 91, 503–527.

Thomas, J.C., McElwain, E.F., Bohnert, H.J., 1992. Convergent induction of osmotic stress—responses. Plant Physiol. 100, 416–423.

Tran, L.S., Urao, T., Qin, F., Maruyama, K., Kakimoto, T., Shinozaki, K., Yamaguchi-Shinozaki, K., et al., 2007. Functional analysis of AHK1/ATHK1 and cytokinin receptor histidine kinases in response to abscisic acid, drought, and salt stress in Arabidopsis. Proc. Natl. Acad. Sci. U. S. A. 104, 20623–20638.

Tsugane, K., Kobayashi, K., Niwa, Y., Ohba, Y., Wada, K., 1999. A recessive Arabidopsis mutant that grows photoautotrophically under salt stress shows enhanced active oxygen detoxification. Plant Cell 11, 1195–1206.

Tyerman, S.D., Skerrett, I.M., 1999. Root ion channels and salinity. Sci. Horticult. 78, 175–235.

Urao, T., Yakubov, B., Satoh, R., Yamaguchi-Shinozaki, K., Seki, M., Hirayama, T., Shinozaki, K., et al., 1999. A transmembrane hybrid-type histidine kinase in Arabidopsis functions as an osmosensor. Plant Cell 11, 1743–1754.

Vaidyanathan, R., Kuruvilla, S., Thomas, G., 1999. Characterization and expression pattern of an abscisic acid and osmotic stress responsive gene from rice. Plant Sci. 140, 21–30.

Verbruggen, N., Hermans, C., 2008. Proline accumulation in plants: a review. Amino Acids 35, 753–759.

Verslues, P.E., et al., 2006. Methods and concepts in quantifying resistance to drought, salt and freezing, abiotic stresses that affect plant water status. Plant J. 45, 523–539.

Wang, B., Luttge, U.L., Ratajczak, R., 2001. Effects of salt treatment and osmotic stress on V-ATPase and V-PPase in leaves of the halophyte *Suaeda salsa*. J. Exp. Bot. 52, 2355–2365.

Wang, Z.Y., Xiong, L., Li, W., Zhu, J.K., Zhu, J., 2011. The plant cuticle is required for osmotic stress regulation of abscisic acid biosynthesis and osmotic stress tolerance in Arabidopsis. Plant Cell 23, 1971–1984.

Wegner, L.H., de Boer, A.H., 1997. Properties of two outward-rectifying channels in root xylem parenchyma cells suggest a role in K^+ homeostasis and long-distance signaling. Plant Physiol. 115, 1707–1719.

Wohlbach, D.J., Quirino, B.F., Sussman, M.R., 2008. Analysis of the Arabidopsis histidine kinase ATHK1 reveals a connection between vegetative osmotic stress sensing and seed maturation. Plant Cell 20, 1101–1117.

Yamaguchi, T., Hamamoto, S., Uozumi, N., 2013. Sodium transport system in plant cells. Front. Plant Sci. 4, 410.

Zhu, J.K., 2003. Regulation of ion homeostasis under salt stress. Curr. Opin. Plant Biol. 6, 441–445.

Microbial ecology of the Namib Desert

6

J.-B. Ramond[1], J. Baxter[1], G. Maggs-Kölling[2], L. Martínez-Alvarez[1], D.A. Read[1], C. León-Sobrino[1], A.J. van der Walt[1], D.A. Cowan[1]

Centre for Microbial Ecology and Genomics, Department of Biochemistry, Genetics and Microbiology, University of Pretoria, Pretoria, South Africa[1] Gobabeb Research and Training Centre, Walvis Bay, Namibia[2]

Chapter outline

1 Introduction

In this book chapter, we focus on the research performed to date on the microbiology and microbial ecology of the Namib Desert, one of the oldest hyperarid deserts on Earth (Seely and Pallett, 2008). Microbial communities consist of complex assemblages of interacting microorganisms, ranging from viruses to small eukaryotes (such as nematodes and springtails) (André et al., 1997; Makhalanyane et al., 2015; Zablocki et al., 2016) and these microbial community members have all been studied, to a certain extent, in the Namib Desert. It is worth noting that most of the

Model Ecosystems in Extreme Environments. https://doi.org/10.1016/B978-0-12-812742-1.00006-4

microbial ecology research performed in the Namib Desert has taken place in the central Namib Desert, in the vicinity of the Gobabeb Research and Training Center (http:/www.gobabebtrc.org/). Consequently, as this desert extends over a latitudinal range of over 2000 km, from the south of Angola to the north of South Africa, there is considerable potential to extend microbial ecology research into the less accessible northern and southern regions. Furthermore, most past microbiological research has been performed on soils and hypoliths, which are microbial communities living on the ventral surface of translucent rocks (Pointing, 2016), while other cryptic and productive desert niches (e.g., endolithic and Biological Soil Crusts; Pointing and Belnap, 2012) have received relatively little attention. It is noted that extensive lichen crust have been detected and mapped in the Namib Desert using remote sensing technologies (Wessels and van Vuuren, 1986; Hinchliffe et al., 2017).

Despite passing mention in a study published in 1960 (Logan, 1960), a further 20 years passed before the Namib Desert microflora were first formally studied. To the best of our knowledge, the culture-dependent study that pioneered microbial ecology in the Namib Desert was a report that noted that fungi were more abundant at the edges of Namib Desert dune Fairy Circles (see Section 6) than in their centers, while anaerobic bacteria displayed the opposite trend (Theron, 1979). Culture-dependent approaches have since largely been replaced by molecular-based fingerprinting and high-throughput sequencing of phylogenetic markers. Nevertheless, in this chapter, we have attempted to review all the scientific literature available on the microbiology and microbial ecology of the Namib Desert.

2 The Namib Desert

The Namib is a low-latitude, African desert that is one of the most arid areas on Earth. It is an isolated, cool, coastal desert of less than 140 km wide along the South Atlantic Ocean shore on the western side of southern Africa (Fig. 1). For the purposes of this chapter, we consider the Namib Desert as an ecological unit, that is, a biome, defined by phytoclimatological characteristics (Irish, 1994; Mendelsohn et al., 2002; Mucina and Rutherford, 2006). In that context, the Namib Desert biome stretches from around Lüderitz in southern Namibia (the Southern Namib) to the furthest extent of the Northern Namibia at Cabo de Santa Maria in Angola (Fig. 1).

The Namib Desert is situated on a coastal peneplain, the Namib Unconformity Surface, which indicates the extensive erosion that occurred after the breakup of the Gondwana supercontinent during the Cretaceous. Oceanographic (Siesser, 1980), paleontological (Senut et al., 2009), and geomorphological (Miller, 2008) evidence indicate a change from subtropical savanna conditions to more arid desert conditions sometime during the Miocene (22–13 Ma [million years before present]). The evidence for the initial onset of desert conditions is contradictory and inconsistent, but there seems to be general agreement that more humid or even wet conditions prevailed during the late Miocene to middle Pliocene (Ward, 1987). Current desert conditions were probably reestablished during the late Pliocene c. 3 Ma, after which

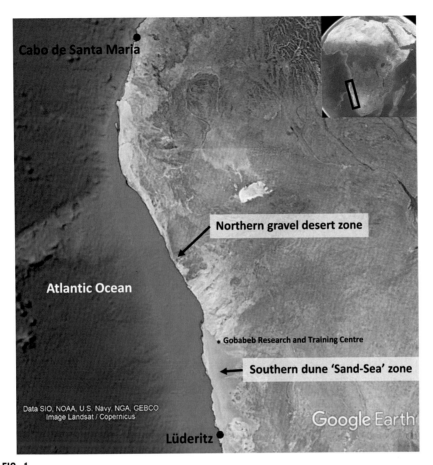

FIG. 1

The Namib Desert. The rectangle on the top right map indicates the localization of the Namib Desert on the African continent.

much of the present-day biota and ecological patterns evolved (Senut et al., 2009). However, even the more recent late Pleistocene paleoenvironmental history of the Namib Desert does not indicate universal arid conditions, but considerable temporal and geographic ecological variation around an arid to hyperarid mean (Stone and Thomas, 2013). The ecological evolution of the Namib Desert should therefore be considered in terms of dynamic extremes.

The principal driver of rapid ecological fluctuations is local hydrology, that is, inputs of moisture. The steep gradient from extreme aridity over the coastal areas of the Namib Desert biome, where almost no rainfall occurs during most years (hyperarid Namib), to the arid conditions over the interior, eastern margin where precipitation ranges from 70 to 100 mm mean annual rainfall, is imposed by the cold coastal waters offshore (Gamble, 1980). The evolution of the Benguela Current

(Siesser, 1980) and the associated Benguela Upwelling System and cold Benguela Coastal Current have a strong influence over hydrological conditions over the adjacent coastal areas. The cold surface conditions over the upwelling cells and coastal current impose quasi-stationary high pressure conditions in the lower atmosphere that impedes the influx of moist air from the interior, while also imposing a temperature inversion in the lower atmosphere that inhibits convection (Eckardt et al., 2013a). The rapid decrease in rainfall isohyets that parallels the coastline reflects the influence of oceanographic conditions over the desert. Rainfall in the Namib Desert biome occurs mainly during the austral summer from convective thunderstorms, but such rainfall events are highly variable, erratic, and localized (Eckardt et al., 2013a). The temporal and spatial distribution of rainfall over the Namib Desert is strongly affected by dynamic fluctuations in upwelling strength, sea surface temperatures, and more distant short-term changes in atmospheric circulation patterns.

In direct contrast to rainfall, fog precipitation and humidity are highest along the coast and rapidly decreases toward the interior (Eckardt et al., 2013a). Fog also seems to be less erratic and more reliable than rain, which have resulted in some unique adaptations in Namib biota to exploit this moisture source. Fog in the Namib Desert results from stratus clouds trapped below the temperature inversion over the cold ocean. Stratus clouds over the coastal areas are advected in over the interior during the late afternoon by thermally induced boundary layer airflow such as the prevalent strong land-sea thermal contrast. Coastal stratus clouds, or "low" fog, are the most common source of near-coast precipitation up to elevations of c. 200 m above mean sea level (amsl), reaching 20–30 km inland and may endure for large parts of the day and night (Eckardt et al., 2013a). Distinct lichen and vegetation communities are associated with this zone of frequent fog precipitation (Juergens et al., 2013). Higher stratus clouds, or "high" fog, may intersect higher elevation areas between 100 and 600 m amsl and up to 60–70 km from the coast. Fog events associated with higher stratus clouds are less frequent and seem to be most common in the austral spring and summer, from September to March. Precipitation from "high" fog usually occurs during the early morning when warmer air, associated with nocturnal airflow from the mountains and plains of the interior, mixes with the stratus clouds (Eckardt et al., 2013a). Various biotic communities in the Namib Desert seem to be predominantly associated with this type of moisture (Juergens et al., 2013).

Other nonrainfall moisture sources such as radiation fog and dew have been described for the Namib Desert (Eckardt et al., 2013a), but have not yet been sufficiently characterized to evaluate their ecohydrological effects.

The opposite gradients of decreasing rainfall from the interior or eastern margin of the Namib Desert biome and a decrease in the duration and effective precipitation from fog moisture, as well as the decrease in the duration and strength of the southerly to southwesterly sea breezes with distance from the coast, have affected pedological processes as well as ecological and evolutionary processes. This is most noticeable in the Central Namib where, in contrast to the Southern Namib Sand Sea, or the complex, fragmentary landscapes of the Northern Namib, most of the desert peneplain is without vast expanses of sand dunes. Not surprisingly, almost all research on the soils of the Namib Desert focused on the Central Namib.

The basement geology of most of the Central Namib peneplain consists of schists, marbles, and quartzites, with occasional inselbergs of granite or ridges of dolerite (Eckardt et al., 2013b). Aeolian erosion and abrasion is more severe closer to the coast where the onshore winds are stronger and more constant (Viles and Goudie, 2013), while fluvial erosion and burial progressively increases toward the eastern margin of the Namib Desert with more rainfall. Most of the peneplains are, however, stabilized by a "desert pavement" consisting of pebbles and coarse sand that protects a thin layer of fine sand (Scholtz, 1972). Outcrops and large rocks are predominantly subjected to mechanical weathering from thermal and aeolian processes, though chemical weathering increases closer to the coast together with more fog precipitation and salt deposition (Viles and Goudie, 2013). Most of the arenosols formed through these processes are "raw" mineral soils with very low organic and clay content (Scholtz, 1972). The most noticeable trait of the Namib Desert soils is, however, the contrasting gradients of gypsum-rich soils (petric gypsisols) that decrease away from the coast, and an observe progressive increase in calcrete soils (petric calcisols) with increased rainfall (Scholtz, 1972). Close to the coast, the stable gypcrete surfaces support abundant lichen crusts and colonies (Loris and Schieferstein, 1992). The striking similarity in the extent of gypsum-rich soils with that of fog moisture, both of which extend 50–60 km inland, is a result of fog-induced gypsum formation and sediments mixed by wind (Eckardt and Spiro, 1999). This entails onshore winds transporting sulfur-rich marine sediments inland, which are then mixed with limestone-rich sediments from the interior that were transported by during the austral winter by strong easterly berg winds, with subsequent wetting of the limestone and sulfur-rich sediments through fog precipitation (Eckardt and Spiro, 1999). The vegetation of the Central Namib clearly responds to these gradients in soils and sources of moisture (Juergens et al., 2013).

The Namib Desert flora is a manifestation of a long evolutionary history in response to aridity. Plant adaptation, diversity, and distribution, as well as population dynamics, in the Namib Desert are driven largely by the spatial and temporal occurrence of precipitation gradients and humidity (Eckardt et al., 2013a). The sparse but relatively diverse plant life has a fairly high proportion of endemic taxa (Juergens et al., 2013). The endemics are associated with a diversity of substrates from sand dunes to gravel plains to rocky inselbergs. Average net primary production throughout this biome is consistently extremely low due to very limited water availability (Mendelsohn et al., 2002).

Characterized by the dominant structure, soil type, and landscape, three discrete vegetation types are defined for the Namib Desert Biome: Northern Desert, Central Desert, and Southern Desert (Mendelsohn et al., 2002). To date, most microbial ecological research has been conducted in the latter two types.

The Central Desert is characterized by sparse shrubs and grasses on the petric gypsisols and petric calcisols of the central-western plains (Mendelsohn et al., 2002). Many shrub species are succulent or long lived to cope with the extreme

FIG. 2

(A) Dune and inter-dune zones in the Namib Sand-sea; (B) Mobile dune crests; (C) Saline desert spring (playa); (D) Namib gravel deserts; (E) Quartz pebble-rich gravel desert pavements; (F) Coastal lichen fields; (G) Hypolithic community; (H) Dune sand "Fairy Circle."

Photo Courtesy: DA Cowan.

and dynamic environment (Juergens et al., 2013). Isotopic analysis of fossil ostrich egg shells showed that C3 plants dominated in the Southern and Central Namib since the Miocene. C4 plants (grasses) only started to proliferate in the Pliocene ca. 3.5 Ma (Ségalen et al., 2006). Generally, the contemporary grass cover is thin and composed of resilient perennial grasses and annual grasses, appearing after sufficient rain has fallen (Henschel et al., 2005). A recent vegetation classification reveals that the Central Desert is composed of 21 diverse units, whose spatial distribution is controlled by climatic gradients, forming a series of vegetation belts that run parallel to the coast line (Juergens et al., 2013). For example, lichens and leaf succulents, many of which are CAM plants, dominate near the coast due to fog precipitation. Plant distribution patterns within these units are further influenced by substrate.

The Southern Desert vegetation type is confined to the Namib Sand Sea landscape (Mendelsohn et al., 2002). The dune sands support specialist plants with exceptional adaptations to unstable substrates, intense abrasion, very low soil fertility, limited moisture, and extreme thermal gradients. A gradual increase in dune vegetation and plant species diversity mirrors the rainfall gradient from west to east (Seely and Pallet, 2008). Some endemic dune species optimize fog as a sustaining water source (Ebner et al., 2011). Individual dunes show a distinctive vegetation gradient across the slope from base to crest (Yeaton, 1988). The inland dunes host a myriad of specialist psammophilous fauna, while the most mobile, coastal dunes are devoid of flora and fauna. Interdune valleys, with a consolidated sandstone and gravel substrate thinly covered by sand, are vegetated by a combination of dwarf shrubs and grasses, but may also be barren. In the interdune grasslands, circular bare patches known as "fairy circles" occur in relatively large numbers (Van Rooyen et al., 2004).

Representation images of the Namib Desert habitats are shown in Fig. 2.

3 **Microbial diversity**
3.1 **Bacteria**

The diversity, colonization, and dispersal of desert/dryland microbial communities have been extensively and recently reviewed (Makhalanyane et al., 2015; Pointing and Belnap, 2012, 2014). The ecological importance of desert microbial communities mainly resides in the fact that, due to the limited presence of higher organisms, they are the main drivers of desert biogeochemical cycling (Makhalanyane et al., 2015; Pointing and Belnap, 2012). This is particularly relevant at a global scale as desert biomes cover a third of the Earth's surface, making them the largest terrestrial biome (Laity, 2009).

The Domain Bacteria has been the most extensively studied in the Namib Desert. Deterministic processes, mainly niche partitioning, have been found to be the major drivers of Namib Desert edaphic bacterial community assembly (Ramond et al., 2014; Ronca et al., 2015; Gombeer et al., 2015; Van der Walt et al., 2016;

Johnson et al., 2017), at both local (e.g., the Kahani Dune; Ronca et al., 2015) and landscape scales (the central Namib Desert; Gombeer et al., 2015; Johnson et al., 2017; Scola et al., 2018). Similar assembly processes were also predicted for hypolithic bacterial communities (Makhalanyane et al., 2013; Stomeo et al., 2013; Valverde et al., 2015), most probably via recruitment of members from the surrounding soils (Makhalanyane et al., 2013).

A range of parameters and variables, including climatic factors (i.e., fog vs rain; Stomeo et al., 2013; Valverde et al., 2015), water regime history (i.e., gravel plain vs riverbed; Frossard et al., 2015), habitat (e.g., soils vs hypolith (Makhalanyane et al., 2013) or open soils versus rhizospheric soils (Valverde et al., 2016)), and soil physico-chemistries (e.g., Gombeer et al., 2015; Ronca et al., 2015) were identified as deterministic forces. However, stochasticity has also been recently identified as an important driver of bacterial community assembly along a 190 km east/west xeric stress gradient in the Namib Desert gravel plains ($n = 80$ sites; Scola et al., 2018).

Using meta-transcriptomics and co-occurrence network analyses, biotic interactions have been strongly implicated in both desert edaphic microbial community assembly processes and diel dynamics (Gunnigle et al., 2017)—a feature already observed in a DNA-based study on Namib Desert hypoliths (Valverde et al., 2015). Co-occurrence network analyses suggested that hypolithic community food webs were driven by a cyanobacterium of the cryptic genus *GpI* (possibly *Anabaena*) (Valverde et al., 2015).

Namib Desert edaphic bacterial communities show similar compositional patterns as in deserts worldwide (Makhalanyane et al., 2015). Whether in the dune sand (Ronca et al., 2015) or the gravel plain soils (Armstrong et al., 2016; Van der Walt et al., 2016), the phyla Proteobacteria and Actinobacteria dominate (Fig. 3). Meta-transcriptomic analyses have also shown that these phyla consistently constituted the majority of the active members of the gravel plain communities (Gunnigle et al., 2017). Metagenomics also revealed that Bacteroidetes was an abundant phylum in in Namib Desert gravel plain soil (Armstrong et al., 2016), as well as in the dune sands (but not, strangely, in interdune soils) (Ronca et al., 2015). It should be noted, however, that Bacteroidetes represented only a minor fraction of the *active* community (Gunnigle et al., 2017). Ronca et al. (2015) also showed that bacterial (sub-) phyla show habitat specificity in the dune habitats. For example, the relative abundances of Chloroflexi, Firmicutes, α-, γ-, and δ-Proteobacteria decreased from dune "top" sites to the interdune zones, while those of Acidobacteria and β-Proteobacteria showed the opposite trend. Dune habitat specificity of some bacterial genera (*Geodermatophilus* sp., *Blastococcus* sp., *Arthrobacter* sp., *Microvirga* sp., *Massilia* sp., and *Novosphingobium* sp.) was also observed (Ronca et al., 2015). Similarly, Cyanobacteria were found more abundant in gravel plain soil communities than in those of dune soils (Fig. 3). These observations appear to confirm the strong influence of environmental filtration in the structuring of Namib Desert edaphic bacterial communities.

A clone library study suggested that Namib Desert hypolithic communities were dominated by the phylum Cyanobacteria; mostly members of the genus

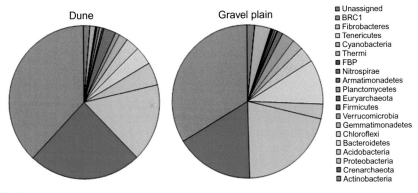

FIG. 3

Average bacterial and archaeal taxonomic composition of Namib Desert dune ($n = 5$) and gravel plain ($n = 5$) Fairy Circle soils. For more information on the generation and the analyses of the 16S rRNA gene sequence datasets, the readers are referred to the Material and Method section of Van der Walt et al. (2016).

Chroococcidiopsis, but with members of the orders Oscillatoriales and Stigonematales also present (Makhalanyane et al., 2013). The dominance of photoautotrophic cyanobacteria in Namib Desert hypoliths was confirmed by pyrosequencing (34% of all sequences; Valverde et al., 2015). This study also showed that the diversity of heterotrophic bacterial taxa was high, with 7 phyla each comprising >3% of the sequences (α-Proteobacteria [22%], Actinobacteria [17%], Acidobacteria [6%], β-Proteobacteria [5%], Bacteroidetes [4%], γ-Proteobacteria [3%], Deinococcus-Thermus [2%]) and an additional 10 phyla with relative abundances of <2%.

Only two studies have, to date, used modern molecular phylogenetics to investigate plant-soil microbiome associations in the Namib Desert. Appropriately, one study focused on the root-associated microbiota of the endemic *Welwitschia mirabilis* "fossil" plant (Valverde et al., 2016). The comparative analysis of five *W. mirabilis* rhizospheric soil communities and open soil communities showed that the latter were significantly different, and more diverse, than the *W. mirabilis* rhizospheric communities. Both communities were dominated by Proteobacteria (particularly, α-Proteobacteria) and Actinobacteria, but showed different dominant genera (Valverde et al., 2016): The rhizosphere communities were dominated by *Nitroreductor* sp. (20%), *Steroidobacter* sp. (12%), *Pseudonocardia* sp. (9%), *Devosia* sp. (4%), and *Glycomyces* sp. (2%), while the open soils communities were dominated by *Rubellimicrobium* sp. (9%), *Kocuria* sp. (5%), *Geodermatophilus* sp. (5%), and *Microvirga* sp. (4%). Only six bacterial OTUs (defined at the 97% similarity cutoff, but representing 38% of the reads) were found in the 5 rhizosphere samples. These phylotypes are thought to represent the *W. mirabilis* rhizosphere core microbiome and were related to well-known plant growth promoting bacteria (*Nitratireductor, Steroidobacter, Pseudonocardia,* and three *Phylobacteriaceae* OTUs). The bacterial communities from the rhizosheath-root system (an adaptive trait of plants from xeric environments) of three Namib Desert speargrasses (*Stipagrostis sabulicola*, *S. seelyae*, and *Cladoraphis spinosa*) were recently investigated (Marasco et al., 2018). This study demonstrated that the rhizosheath-root system communities were recruited from the surrounding sandy soils, with strong compartment (i.e., rhizosphere, rhizosheath, and root) niche-partitioning and independently from the plant species. Furthermore, they were dominated by Actinobacteria and Alphaproteobacteria (e.g., *Lechevalieria*, *Streptomyces*, and *Microvirga*). Given the high degree of plant endemism in the Namib Desert (Seely and Pallett, 2008), there is considerable scope for further development of research on plant-microbe interactions in the Namib. Such studies would most certainly lead to a better understanding of plant-microbe interactions in hot and arid environments, and could be fundamental in predicting plant (including food crop) adaptation in a global warming scenario.

With a focus on community dynamics, a year-long soil microbiome study in the gravel plains showed that Namib Desert bacterial community structures were largely static over time, but that a rainfall event triggered significant structural and functional changes (Armstrong et al., 2016); this result is consistent with the proposals that desert ecosystems are water pulse driven at the microbial scale (Belnap et al., 2005). A month after the precipitation event, bacterial communities had nearly recovered their initial prerainfall compositional status (Armstrong et al., 2016). Over much shorter timescales, a meta-transcriptomics analysis has contradicted the general belief in desert microbial communities are inactive except during water pulses. The functional community was shown to show substantial short-term dynamism, fluctuating significantly over five diel cycles during a dry period (Gunnigle et al., 2017). The identification of 110 bacterial and archaeal proteins in a Namib Desert gravel plain soil further supports the suggestion that desert soil communities are active even under dry conditions (Gunnigle et al., 2014).

3.2 Fungi

Molecular-based analyses of Namib Desert fungal communities have only recently been reported (Ramond et al., 2014; Van der Walt et al., 2016; Valverde et al., 2016; Johnson et al., 2017). Deterministic niche partitioning was identified as the key process involved in Namib Desert edaphic fungal community assembly, independent of the molecular technique used (fingerprinting [Ramond et al., 2014; Johnson et al., 2017] or barcoding [Van der Walt et al., 2016]), in that each Namib Desert habitat studied was found to harbor a specific fungal community. However, it should be noted that when using T-RFLP fingerprinting, no fungal signatures were detected in dune "top" and "slope" zones (Johnson et al., 2017). Conversely, Illumina MiSeq sequencing showed 553 ($\pm$88) and 366 ($\pm$59) fungal OTUs in vegetated dune soils ($n = 5$) and in dune Fairy Circle soils ($n = 5$), respectively. Environmental filters identified as drivers of Namib Desert fungal communities structures were NH_4^+-N content, Cation Exchange Content, %C (total), and %Clay in gravel plain soils and wet riverbed soils, while %sand was a significant factor in the structuring of

Saltpan, Dry Riverbed, and Interdune zone fungal communities (Johnson et al., 2017). Another recent study comparing Namib Desert dune and gravel plain edaphic fungal communities showed that the dune communities were significantly influenced by %sand, pH, and calcium content, while the gravel plain communities were driven by soil phosphorous, sodium, sulfur, and carbon content and %silt and %clay (Van der Walt et al., 2016).

The phylotypic composition of Namib Desert soil fungal communities was dominated by Ascomycota (c.f., Basidiomycota) (Van der Walt et al., 2016; Valverde et al., 2016; Marasco et al., 2018). Fungi of the class Dothideomycetes were found to be significantly more abundant in gravel plain soils than in dune soils, while the classes Chytridiomycota, Sordariomycetes, and Agaricostilbomycetes showed the opposite trend (Fig. 4). Valverde et al. (2016) also noted the dominance of Dothideomycetes (in decreasing order, genera *Aspergillus*, *Spiromastix*, *Phoma*, and *Alternaria*) in the gravel plain soils while Eurotiomycetes and Sordariomycetes dominated the *W. mirabilis* rhizospheric soils (*Alternaria* and *Lecythophora* genera; Valverde et al., 2016). An interesting (and potentially concerning) observation was that, in 8 sites spanning the northern Namib Desert (i.e., from the central Namib Desert [24°S] to the Angolan boarder [18°S]), the phytopathogenic fungus *Aspergillus niger* var. *phoencis* infected 82% ($n = 616$) of female *W. mirabilis* plants (Cooper-Driver et al., 2000). In contrast, vesicular-arbuscular mycorrhizal fungi of the genus *Glomus* were associated with *W. mirabilis* rootlets in four of seven sites studied: in only 2 did all the plants investigated show mycorrhizal structures (Jacobson et al., 2003). The authors concluded that the colonization of *W. mirabilis* rootlets by mycorrhizal fungi was highly dependent on rainfall, and grass distribution and abundance, both of which are also rainfall dependent (Jacobson et al., 2003).

Culture-dependent studies of Namib Desert soil samples have demonstrated that fungal species are important decomposers, matching detritivores and abiotic factors

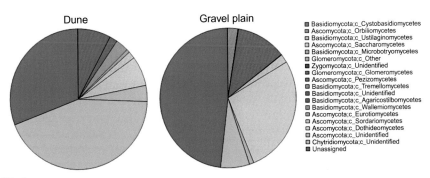

FIG. 4

Average fungal taxonomic composition of Namib Desert dune ($n = 5$) and gravel plain ($n = 5$) Fairy Circle soils. For more information on the generation and the analyses of the ITS region sequence datasets, the readers are referred to the Material and Method section of Van der Walt et al. (2016).

(Jacobson, 1997; Jacobson et al., 2015). Moisture, in the form of fog or rain, was found to activate fungal-dependent decomposition in the Namib dunes (Jacobson and Jacobson, 1998; Jacobson et al., 2015). At 11 sites spanning two longitudinal transects in the Namib Sand-Sea, 12 mycorrhizal fungi were found to be associated with a wide variety of plants from the genus *Stipagrostis* sp. (*S. sabulicola, S. lutescens, S. seelyae, S. ciliata*), and with *Centropodia glauca* and *Cladoraphis spinosa* (Jacobson, 1997). Spores from five *Glomus* species dominated all samples, representing over 95% of the spores collected (Jacobson, 1997). Fungal spore diversity and abundance in both transects was shown to be linked to sand stability (i.e., spore counts increased from the mobile dune "top" zones to the stable interdune zones). Moisture availability also appeared to significantly impact mycorrhizal colonization (Jacobson, 1997).

Twelve *Glomus* species were associated with different Namib Desert open and rhizospheric soils with *G. aggregatum* being the most prevalent (Uhlmann et al., 2006). Stutz et al. (2000) also isolated numerous arbuscular mycorrhizal fungi of the *Glomus* genus from Namib Desert soils (*G. eburneum, G. etunicatum, G. intraradices, G. microaggregatum, G. mosseae, G. occultum, G. spurcum, G.* AZ112, *G.* AZ123, *G.* TX106 and *G.* NB118) as well as the genospecies *Acaulospora morrowiae* and *Acaulospora trappei*. These authors noted that the Namib and North American desert soils shared between 54% and 79% of arbuscular mycorrhizal fungi species compositions, suggesting that extensive dispersal processes, possibly coupled with common deterministic environmental parameters, may be involved in the structuring of desert fungal communities (Stutz et al., 2000).

3.3 Archaea

Over recent years, metagenomic analyses of several Namib Desert niches (saline playas, hypolithic communities, and open soil environments) have demonstrated the presence of significant archaeal communities, although detailed compositional surveys are largely unavailable. In saline playas, for example, T-RFLP analyses suggested that archaea are the most diverse prokaryotes, their diversity being shaped strongly by ion concentrations, carbon content, and soil silt and moisture (Johnson et al., 2017). The detection of a small number of viral contigs resembling those infecting haloarchaea of the genus *Haloarcula* suggests that haloarchaea may be dominant in these communities (Adriaenssens et al., 2016). A more detailed analysis of the distribution (and importance) of archaea in the total saline soil communities is still missing. In hypolithic microbial communities, Makhalanyane and colleagues initially reported the presence of the methanogenic euryarchaeon *Methanococcus vanniei* and the hyperthermophilic crenarchaeon *Sulfolobus acidocaldarius* (Makhalanyane et al., 2013). More recent studies showed that only 0.38%–0.43% of the hypolithic community phylotypes were Archaeal (Le et al., 2016; Vikram et al., 2016). Archaeal contigs from a whole metagenome analysis demonstrate a clear dominance of the Euryarchaeota (57.5%) over the Thaumarchaeota (34.6%)

and Crenarchaeota (7.9%), although (surprisingly) genes indicating the presence of methanogens could not be detected (Vikram et al., 2016).

In contrast to the Euryarchaeal-dominated hypolithic communities, Thaumarchaea is considered to be the dominant archaeal phylum in global hot desert soils, from 1% to 3% (Bates et al., 2011) to 6.6% of total phylotypic sequences in North American desert soils (Fierer et al., 2012). Only 2% of Crenarchaeal/Thaumarchaeal OTUs were observed in Namib gravel plain soil sample in a year-long study, although their relative abundance doubled after a single rainfall event (Armstrong et al., 2016). Another recent study of Namib dune and gravel plain soils, together with soils from within Fairy Circle formations, reported that although Archaeal OTUs represented only about 2% of the total phylotypes, the contribution of Crenarchaea/Thaumarchaea to the prokaryotic communities ranged from 10% to 25% of the total (Fig. 3: Van der Walt et al., 2016). Most of the Archaeal phylotypes were found in both dune sand and gravel soil samples, suggesting the existence of a broad core community of archaea in bulk hyperarid soils. Only 18% of the archaeal phylotypes detected were specific to either of the two soil types, in similar proportions. Interestingly, an ammonium-oxidizing Thaumarchaeal OTU was detected (uniquely) in the interior of the Fairy Circles in both soil types. These results would place Namib desert soils among the most archaea rich of all global desert soils (Bates et al., 2011). Archaeal community diversity in these nonsaline open soils is driven by several abiotic factors similar to those shaping bacterial diversity, but analyses of soil samples from gravel plains suggest that NH^{4+} and Mg^{2+} specifically affect archaea in gravel plain and dune soils (Van der Walt et al., 2016), although only the former could be confirmed to be a significant contributor in a posterior T-RFLP OTU diversity study (Johnson et al., 2017). This important role of NH^{4+} would be consistent with the dominant presence of ammonia-oxidizing thaumarchaeal groups in these soils.

Contrasting with DNA-based surveys consistently reporting a community with minor euryarchaeal components, a proteomic analysis of soils from a gravel plain site detected 39 euryarchaeal proteins classified mostly as having methanogen origin, whereas only 12 different crenarchaeal proteins could be identified. An abundance of DNA replication and repair proteins from both groups suggests that archaea are active members of the microbial community (Gunnigle et al., 2014).

3.4 Viruses

Viruses associated to microbial communities are regarded as an important factor in dictating taxonomic and functional diversity of the community (Kimura et al., 2008; Srinivasiah et al., 2008; Rohwer and Thurber, 2009; Brum and Sullivan, 2015). Although the ecological role of viruses in desert soils is still poorly understood, considerable effort has been directed in recent years to the study of the viral diversity associated to the microbial communities of the Namib Desert, mainly through the employment of Next Generation Sequencing methods.

DNA metaviromes of hypolith, surface soil (including copper-rich soil) and solar saltern microbial communities of the Namib Desert have been very recently analyzed. These datasets have a high proportion of sequence reads with no homology to any sequences deposited in the public databases (76%–95%) (Adriaenssens et al., 2015, 2016; Zablocki et al., 2017; Hesse et al., 2017), highlighting the novelty of the Namib Desert metaviromes.

The analyses of the (mostly dsDNA) viral fractions of the different Namib Desert niches sampled (with exception of the solar salterns) show that the viral taxonomic composition is dominated by members of the order *Caudovirales*, with the family *Siphoviridae* being the most abundant, followed by the *Myoviridae* and *Podoviridae* families. This is in agreement with transmission electron microscopy analyses of the viral fractions of hypoliths and surface soils that show a marked abundance of morphotypes belonging to the *Siphoviridae* and *Myoviridae* families (Prestel et al., 2008; Adriaenssens et al., 2015; Hesse et al., 2017).

The Namib-hypolith metavirome shows a prevalence of viruses of the order *Caudovirales*, with a dominance of unclassified *Geobacillus*- and *Bacillus*-infecting phages, as well as ssDNA bacterial viruses of the *Microviridae* family (Adriaenssens et al., 2015). Only a few sequences with distant similarity to database cyanophage sequences were identified (Adriaenssens et al., 2015), despite the evidence that the cyanobacteria represent one of the most abundant phyla in the Namib Desert hypolithic communities (Vikram et al., 2016). This is attributed to the fact that cyanophage sequences deposited in public sequence databases are almost exclusively of marine origin, and suggests that desert soil cyanophages are only very distantly related to the marine cyanophages (Adriaenssens et al., 2015).

In comparison, the metaviromes from surface soils obtained along a 120 km longitudinal transect of the Namib Desert show a predominance of *Mycobacterium*- and *Escherichia*-infecting viruses, followed by *Rhodococcus*-, *Bacillus*-, and *Geobacillus*-infecting phages, all members of different families of the order *Caudovirales* (Zablocki et al., 2017; Scola et al., 2018). A copper-rich Namib soil metavirome, obtained from a coastal fog influenced site, has been recently published, in which the authors use a highly stringent approach for the taxonomic annotation of viral open-reading frames (ORFs). Members of the *Caudovirales* dominated in the sample and, among these, several ORFs showed similarity to phages infecting the genera *Pelagibacter*, *Thalassomonas*, *Pseudoalteromonas*, and *Cellulophaga*, all marine-associated bacteria. Hits to *Lactococcus*- and *Azospirillum*-infecting phages suggest a link to rhizobial symbionts. Notably, archaeal haloviruses dominated in the fraction of unclassified dsDNA viruses. The sequence fraction corresponding to ssDNA viruses revealed novel circovirus-like contigs that are related to circoviruses of aqueous and marine-associated environments (Hesse et al., 2017). The authors hypothesized that the multiple sequence hits in the copper-soil metavirome to viruses of marine environments could arise from microorganisms (and their associated viruses) that are transported by fog and wind to the sampling site, due to its proximity to the coast (Hesse et al., 2017). This suggestion is questionable, however, as similar marine sequence signals were not observed in other soil samples recovered from the fog zone at less distance from the coast (Zablocki et al., 2017).

In contrast to the soil and soil niche habitat metaviromes, the metaviromes obtained from two solar salterns in the Namib Desert showed a range of novel ssDNA viruses and caudoviruses. The metavirome of the Eisfeld playa, located near Swakopmund, showed numerous ssDNA viruses of the bacteria-infecting *Gokushovirinae* subfamily (*Microviridae* family) and several ssDNA circoviruses. Some contigs were related to halobacteria- and cyanobacteria-specific viruses, suggesting the presence of phages that infect these microorganisms. The metavirome from the Hosabes playa, near Gobabeb showed dsDNA viruses of the *Myoviridae* family and ssDNA phages of the *Microviridae* family and its *Gokushovirinae* subfamily, the *Inoviridae* family and the plant-infecting *Geminiviridae* family. Viruses distantly related to the dsDNA *Salterprovirus* genus were also identified in this metavirome, suggesting that these viruses may represent a novel lineage of haloarchaeophages. There was considerable similarity in the viral community structures of the two solar salterns studied, despite the geographical distance between them (approximately 124 km), with over 55% overall sequence similarity between the two metaviromes (Adriaenssens et al., 2016). The similarity in the viral community composition of both playas was attributed to the similarities of the two environments, that is, shallow, salt-rich springs (Adriaenssens et al., 2016).

Interestingly, the presence of large dsDNA virus genomes, belonging to the proposed order *Megavirales*, has been identified in several samples from the Namib Desert. A limited number of hits to the algae-infecting *Phycodnaviridae* family and to the amoeba-infecting *Mimiviridae* family have been reported in a Namib hypolith metagenome and metavirome (Vikram et al., 2016; Adriaenssens et al., 2015), in the Namib transect soil metaviromes (Zablocki et al., 2017; Scola et al., 2018) and in the Swakopmund Eisfeld saltern metavirome (Adriaenssens et al., 2016).

As found in other desert ecosystems (Zablocki et al., 2016), a high prevalence of lysogenic viruses is expected in the Namib Desert viral community. Indeed, an early study of the viruses obtained from surface soils from the Namib Desert reported an increase in the abundance of viral DNA genomes detected by pulse-field gel electrophoresis after treatment with mitomycin C, a DNA-damaging agent commonly used for the induction of prophages (Prestel et al., 2008). The identification of multiple phage integrase genes in the Namib solar saltern metaviromes also suggests the presence of numerous lysogenic viruses (Adriaenssens et al., 2016). It is hypothesized that the lysogenic mode of infection is preferred under conditions that might induce the degradation of viral particles (such as high temperatures, desiccation, and high UV irradiation fluxes) and where the diffusion rate of viral particles is expected to be low, such as in hyperarid desert soils (Zablocki et al., 2016).

Despite recent efforts to describe the viral diversity of the microbial communities in the Namib Desert, very little is known about the virus–host relationships, life cycles, viral population dynamics, or the functional role of the viruses in this ecosystem. Future research efforts should be aimed at addressing these questions.

3.5 Lower eukaryotes

Lower eukaryotes include protozoa, slime molds, algae, and other single- and multi-cellular eukaryotes (Wickner, 2012). Microfauna such as nematodes and microarthropods may also be included in this group. It has been long established that the Namib Desert, with its unique geomorphology and climatic conditions (Seely, 1990), is associated with a wide range of unique flora and fauna (Nicholson, 2011), and it might be expected that the arthropod populations should show a similar diversity and novelty. Indeed, highly diverse communities of ciliates (protozoa) have been found in terrestrial and semiterrestrial habitats in the Namib (Foissner et al., 2002). Lichens, symbiotic associations of green algae and fungi, have extensively colonized quartz and soil substrates in the Namib (Lalley et al., 2006; Büdel, 2011). The morphological characterization and distribution in the Namib Desert of the chasmoendolithic *Lecidea*, which colonizes cracks in quartz rock has been the focus of detailed research (Wessels, 1989) The distribution of lichens is heavily dependent on fog-water input, and largely restricted to the coastal fog zone (Schieferstein and Loris, 1992). *Parmelia hueana* Gyeln., for example, is a lichen species solely found in the fog-influenced coastal zone of the Namib Desert (Büdel and Wessels, 1986), where lichen "fields" (regions where lichens cover over 20% of the desert pavement surface area) are numerous and highly species rich (averaging 12 lichen species per square meter; Schieferstein and Loris, 1992).

While the Namib Desert has been the focus of much microbiological research, few studies have focused on the ecology of Namib soil arthropods, compared to North American deserts, for example. Marsh (1987) compared the structure of microarthropod communities from open soils, *W. mirabilis* litter and rhizospheric soils originating from different sites across the Namib Desert. Globally, rhizospheric soils were significantly richer in microarthropods than open soils but community structures varied between sites. The northern Namib communities were significantly less dense (1710 ± 720 microarthropods m^{-2}) than in the southern Namib ($145,700 \pm 95,930\ m^{-2}$). André and colleagues (André et al., 1997) isolated a total of 183 mites, 171 collembolans, and 47 microarthropod species from gravel plain samples (under *W. mirabilis* and *S. ciliata* plants), dune sand (under *Acanthosicyos horridus* [!Nara] and *S. sabulicola*), and riverbed soils. The arthropod and mite richness was strongly correlated with soil potassium and sodium content, as well as C/N ratios. This study was only performed over a period of 2 weeks and in very localized areas, highlighting the enormous gaps in the scientific knowledge of Namib Desert microarthropod diversity, abundance, and distribution.

Lichens and associated biological soil crusts are important components of ecosystems in arid and semiarid landscapes. Along the Namib Desert coastal fog zone, extensive lichen fields are thought to make an important contribution as primary producers (i.e., via C and N fixation). These soil crust communities also play an important role in regulating soil temperature, soil moisture, and contribute to the prevention of soil erosion (Le Roux, 1970; Lange et al., 1994). Heterotrophic Protista have been studied in a wide range of habitats on numerous continents, but biological soil crusts in southern Africa have not been as intensively studied. Foissner et al. (2002) identified 365 ciliate species across Namibia (including from the Namib Desert), 35% of which had not previously been described. It seems reasonable to assume that Namib Desert niche habitats (biological soil crusts, hypoliths, etc.) may harbor an equal diversity of hitherto undiscovered protist species.

There is huge potential for further study of lower eukaryote diversity in the Namib Desert. Current research using modern molecular fingerprinting tools has shown an astonishing genetic diversity of arthropods (e.g., Springtails (Collembola): Baxter, Cowan & Hogg, unpublished results) in Namib Desert soils.

4 Drivers of microbial community structure

Deserts are water pulse-system driven environments (Noy-Meir, 1973; Belnap et al., 2005). Despite regions of hyperaridity, the Namib Desert is unusual in that it is characterized by a well-defined longitudinal xeric stress gradient with regular fog events penetrating up to 75 km inland and increasing rainfalls from the eastern (inland) mountains (Eckardt et al., 2013a). This xeric gradient has led to a high level of faunal and floral endemism (Seely and Pallett, 2008). Moisture source (i.e., fog vs rain) has also been found to influence Namib Desert edaphic and hypolithic community assemblages (Stomeo et al., 2013; Valverde et al., 2015; Scola et al., 2018), hypolithic colonization processes (Warren-Rhodes et al., 2013), soil extracellular enzyme activities (Scola et al., 2018), fungal decomposition rates (Jacobson et al., 2015), and the physiology (specifically, photosynthesis and CO_2 exchange rates) of the lichens *Acarospora* cf. *schleicheri*, *Caloplaca volkii*, *Lecidella crystallina*, and *Telschistes capensis* (Lange et al., 1994, 2006). For example, 100% of the quartz pebbles in the Namib Desert gravel pavement were colonized in zones where mean annual precipitation (MAP) reached 40–60 mm. In the fog zone, where MAP is ~25 mm, the 100% colonization rate suggests that fog water is an important source of precipitation and driver of hypolith community development (Warren-Rhodes et al., 2013).

Soil and hypolithic microbial community structures were found to be significantly different between the fog and rain zones of the Namib Desert (Scola et al., 2018; Valverde et al., 2015). The consistently higher salt content in the fog zone soils was identified as an important environmental filter of microbial community structures and functional capacities (Scola et al., 2018; Stomeo et al., 2013). Long-term climatic parameters (i.e., fog vs rain as the principal source of water input) have been shown to influence Namib Desert soil microbial activities across a 190 km longitudinal transect (Scola et al., 2018), where a significant and positive relationship was observed between the "distance-to-coast" of the sampling sites ($n = 20$; significantly correlated to mean annual precipitation) and the activities of five extracellular enzymes used as proxies of edaphic C, N, or P biogeochemical cycling processes (Frossard et al., 2012), that is, fluorescein diacetate hydrolysis (FDA), β-glucosidase (BG), alkaline phosphatase (AP), leucine aminopeptidase (LAP), and phenol oxidase (PO) activities.

A microcosm study testing the effect of different wetting events (i.e., mimicking either fog, light-rain or high-rain intensities and frequencies) on Namib Desert soils with contrasted water regime histories (riverbed vs gravel plain) confirmed that soil extracellular enzyme activities were soil specific (Frossard et al., 2015). BG and LAP activities were, for example, significantly higher in the gravel plain than in riverbed soils. Furthermore, extracellular enzyme activities did not show clear trends after the water inputs. The bacterial and fungal communities of the different soils were also initially significantly different (as measured by T-RFLP fingerprinting). However, while gravel plain fungal communities consistently demonstrated resilience toward the different watering treatments, the riverbed communities were significantly modified. This study also demonstrated that intensity (i.e., the total amount of water added) rather than the frequency of the wetting events had the greatest impact on Namib Desert soil microbial communities.

In the context of global climate change, we suggest that such controlled multifactorial experiments, for example, testing the impact of intense water addition $\pm$ temperature increase $\pm$ elevated atmospheric CO_2 concentrations, can provide valuable information on the structural and functional responses of different soil microbial communities to different climate change variables. Given that hot desert biomes are considered to have the potential for greater climate change impact than most other terrestrial biomes (Schlesinger et al., 1990) and given their importance on a global scale (constituting around a third of the Earth's surface; Pointing and Belnap, 2012), we argue that the addition of microbially mediated variables to current climate change models (Docherty and Gutknecht, 2012) would add depth and value to these models.

5 Adaptive mechanisms in desert microbes

Hyperarid desert environments, such as the Namib, are characterized by high xeric stress, extremes in temperature and ultraviolet radiation, and low levels of available nutrients (Pointing and Belnap, 2012). The stress-related adaptations of Namib Desert plants have been well documented (Juergens et al., 2013; Van Heerden et al., 2007) but much less is understood regarding microbial adaptations in the same environment (Le et al., 2016). Nevertheless, the evidence of substantial microbial populations of high species diversity in these xeric soils suggests that the extant microorganisms have comprehensively adapted to the stresses imposed by these "extreme" environmental conditions (Lebre et al., 2017). The nature of these adaptations may be behavioral, physiological, or structural at macroscopic, microscopic, or molecular levels. Such adaptations include, for example, a hypolithic or endolithic lifestyle, the excretion of lipopolysaccharides, dormancy and endospore formation (Majid et al., 2016), membrane modifications and solute accumulation (Ranawat and Rawat, 2017) or molecular adaptations, such as the expression of chaperone and heat-shock proteins (Rajeev et al., 2013).

Stress avoidance, through the colonization of niche habitats, is common among Namib Desert microbes. These refuge environments moderate and shield communities from harsh environmental stressors such as UV radiation, xeric stress, and temperature extremes. These niches range from the widespread, more macroscopic biological soil crusts (Belnap, 2003), to more cryptic lithic niches such as epi- and hypoliths, which inhabit rock surfaces (Chan et al., 2012; Pointing and Belnap, 2012) and endoliths which colonize the porous spaces within rocks (Walker and Pace, 2007). The overlapping fog/rainfall precipitation regimes in the Namib Desert has resulted in an almost uniform distribution of hypolithic communities (Eckardt et al., 2013b), which have made them the most extensively studied niche habitats in the Namib Desert. Like epiliths, hypoliths colonize the outer surfaces of rocks: however, hypoliths occupy an even more cryptic refuge, being found exclusively on the ventral side of translucent pebbles (Chan et al., 2012). The hypolithic environment allows for increased protection against UV radiation and desiccation, while still providing sufficient photosynthetically useful light for primary production, by predominantly cyanobacterial species (Makhalanyane et al., 2013).

Physiological adaptations of Namib microbes allow for their survival over periods of direct exposure to extreme environmental conditions. Sporulation and metabolic slowdown is a major adaptive strategy among desert microbes, especially for members of the phylum Actinobacteria, which are common in Namib Desert soil communities (Makhalanyane et al., 2015). The production of secondary metabolites is another strategy utilized by extremophiles (Makhalanyane et al., 2015). Cyanobacteria, which are usually the dominant primary producers in Namib Desert soil ecosystems, particularly soil niche habitats (Valverde et al., 2015), have a number of unique adaptations for offsetting the effects of high UV radiation. These include the production of UV-absorbing compounds, such as scytonemin, and highly efficient mechanisms for photosystem repair (Makhalanyane et al., 2015). Bacterial genera such as *Deinococcus*, which are capable of withstanding high levels of radiation, have also been observed in Namib Desert microbial communities (Ronca et al., 2015). Arid and hyperarid habitats are usually associated with high salt and osmotic stress. High ion concentrations can lead to the damage and inactivation of cellular machinery, such as the cyanobacterial photosystems (Makhalanyane et al., 2015). Halotolerant microbes moderate osmotic stress either by accumulating similar cytoplasmic ion concentrations as the environment, using influx systems, or by moderating osmotic balance through the accumulation of organic osmolytes (Yaakop et al., 2016).

With the advent of sequencing-based technologies and more recently "omics-based analyses," it has been found that these communities rely on complex taxonomic and functional interactions. Metagenomic analyses have been used to estimate the diversity and functional capacity of Namib Desert microbial communities (Adriaenssens et al., 2015; Ronca et al., 2015; Vikram et al., 2016). Le et al. (2016) made use of metagenomics analyses to determine which stress-related genes are associated with Namib Desert hypolith communities. Among these were kinases involved in temperature regulation, gene families involved in oxidative stress limitation, and transport systems for the uptake of protective osmolytes.

Due to the limited diversity of higher eukaryotic primary producers in hot deserts, such as the Namib, microorganisms are responsible for carrying out essential ecosystem services such as carbon and nitrogen cycling. Any losses of microbial species that perform nonredundant functions could have a negative impact on the entire biome (Makhalanyane et al., 2015). It will therefore be important to perform more research on how microbial communities adapt to changes such as desertification, especially since desert ecosystems are predicted to expand due to continuing climate change (Pointing and Belnap, 2012).

6 The Fairy Circle enigma

Desert environments typically exhibit vegetation patterning (Rietkerk et al., 2004). One of the most well-known examples, the enigmatic Fairy Circles (FCs), can be found in the Namib Desert. Fairy Circles are circular soil patches completely devoid of any plants, surrounded by a peripheral rim of taller grass within a matrix of vegetation (Fig. 5). They are typically between 2 and 10 m in diameter and have an average "life-span" of between 22 and 60 years (Tschinkel, 2012; Albrecht et al., 2001). FCs are found in the coastal regions of Namibia (60–120 km inland) which receive between 50 and 150 mm precipitation per annum (Van Rooyen et al., 2004; Tschinkel, 2012). FCs in both the Namib Desert dune fields and gravel plains have been well documented (Cox, 1987; Van der Walt et al., 2016; Ramond et al., 2014, and references therein), but have only recently been discovered elsewhere (the Australian outback: Getzin et al., 2016). However, despite being the topic of vigorous research for more than 40 years, their origins and "physiology" remain unexplained.

Early hypotheses on the origin of FCs suggested that they were either remnants of ancient termite nests (Tinley, 1971), due to plant allelopathy (Theron, 1979), the result of hydrocarbon gas seepage (Naudé et al., 2011), or created by microfauna

(A)   (B)

FIG. 5

Fairy Circles of the Namib Desert dune fields (A) and gravel plains (B).

Photographs courtesy of D.A. Cowan and J.-B. Ramond.

(e.g., Albrecht et al., 2001). While many of these possible causes have subsequently been refuted (Van Rooyen et al., 2004; Becker and Getzin, 2000; Tschinkel, 2012), FCs have recently received increased attention. The most prominent and still valid hypotheses include microfaunal ecosystem engineering (e.g., Juergens, 2013) and vegetation self-organization (Getzin et al., 2016). Tarnita et al. (2017) recently suggested that FCs could have a multifactorial origin (as proposed by most previous investigators), that is, a result of the interplay between termite colony formation and extreme aridity which results in the limited vegetation growth seen within FCs.

An alternative hypothesis, originally proposed by Eicker et al. (1982), is that FCs could be caused, or maintained, by phytopathogenic edaphic microorganisms. This theory is supported by findings from Van Rooyen et al. (2004), which showed that while grass seedlings are able to germinate the barren patches of FCs after rainfall events, the seedlings do not survive long after germination. These authors also demonstrated that soil from within FCs inhibited plant growth and root development, strongly implicating a phytotoxic mode of action.

As corroborative (albeit circumstantial) evidence, many natural circular phenomena result from the action of microorganisms. Fungi, in particular, are responsible for various circular necrotic phenomena. The causative agents for turf grass fairy rings are the fungi *Vascellum curtisii* and *Bovista dermoxantha* (Fidanza et al., 2016; Terashima et al., 2004). Within intertidal mats, the circular clearings in cyanobacterial and diatom microbial mats have been shown to be the result of the fungal *Emericellopsis* sp. (Carreira et al., 2015). Dollar spot of turf grass is a fungal infection by *Sclerotinia* sp. (Walsh et al., 1999), while blighting of Antarctic microbial mats has recently been shown to be the result of high fungal abundance (Velázquez et al., 2016). The fungal origins of such "circular phenomena" have stimulated interest in edaphic microbial pathogenesis as a causative agent of the FC phenomenon.

Ramond et al. (2014) found both fungal and bacterial communities to be significantly different between Namib Desert gravel plain FC soils and vegetated control soils using T-RFLP community fingerprinting. These findings have recently been corroborated (Van der Walt et al., 2016) using Illumina MiSeq barcoding of both bacterial and archaeal (16S rRNA gene) and fungal (ITS region) phylogenetic markers, showing that FC microbial communities were significantly distinct from those found in the surrounding vegetated control soils. In addition, such differences in community structure were observed in both the environments in which fairy circles are found in the Namib Desert: the gravel plains and dune fields.

Fungi in the Namib Desert FC environments were also found to be zone specific, that is, significantly different between dunes and gravel plains, as well as between control and FC center soils (Van der Walt et al., 2016). Dune and gravel plain Fairy Circle bacteria/archaeal communities were dominated by the Actinobacteria, Proteobacteria, and Crenarchaeota phyla (Fig. 6). Globally, fungi of the class Chytridiomycota and Wallemiomycetes were significantly more abundant in dune soils, while gravel plain fungal communities consistently contained more Dothideomycetes (Fig. 7). Van der Walt et al. (2016) identified 57 fungal phylotypes unique to FC soils (i.e., not present in vegetated control soils), where several phylotypes could be

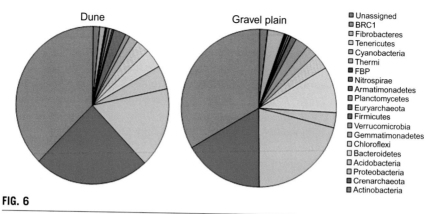

FIG. 6

Average bacterial and archaeal taxonomic composition of Namib Desert dune ($n = 5$) and gravel plain ($n = 5$) Fairy Circle soils. For more information on the generation and the analyses of the 16S rRNA gene sequence datasets, the readers are referred to the Material and Method section of Van der Walt et al. (2016).

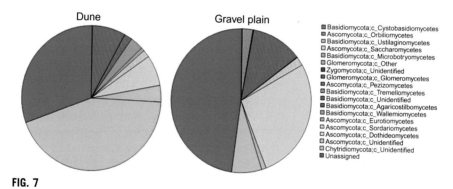

FIG. 7

Average fungal taxonomic composition of Namib Desert dune ($n = 5$) and gravel plain ($n = 5$) Fairy Circle soils. For more information on the generation and the analyses of the ITS region sequence datasets, the readers are referred to the Material and Method section of Van der Walt et al. (2016).

assigned to well-known fungal phytopathogenic taxa (e.g., genera *Periconia*, *Curvularia*, and *Aspergillus*, order Pleosporales and family Chaetomiaceae (Falloon, 1976; Hedayati et al., 2007; Zhang et al., 2012; Viola et al., 2007)). While clearly not proof of causality, these findings are completely consistent with the view that microorganisms may be closely involved in the formation or maintenance of FCs.

In the opinion of the authors, the Fairy Circle phenomenon remains currently unexplained, on the grounds that no single hypothesis satisfactorily incorporates all available observations and features of FCs in explaining their etiology. Such "anomalous" features include, for example: the wide geographic distribution of

FCs (Namib dune, gravel plain and Australia), the plant growth inhibitory capacity of FC soils, the coexistence of both regular and highly irregular distributions, and the differences observed in both microfaunal and microorganism distribution between FC and control soils.

7 The Namib Desert, an understudied astrobiology model system

Extremophiles are (micro)organisms, essentially from the Bacteria and Archaea domains, thriving under the most extreme environmental conditions. Consequently, their adaptive strategies and physiological capacities are widely used to design hypotheses and build theories on how and where Life could develop on extraterrestrial planets or moons (Cavicchioli, 2002; Fairén et al., 2010). Due to their polyextreme nature, with notably high UV, hot or cold temperatures and desiccation stresses, hyperarid deserts are often regarded as ideal model systems to perform astrobiology studies (Fairén et al., 2010; Billi et al., 2013).

While many of these have stemmed from the hot Atacama Desert and the cold Antarctica Dry Valleys (e.g., Gilichinsky et al., 2007; Fairén et al., 2010; Parro et al., 2011; Billi et al., 2013), surprisingly almost none have originated from the hyperarid Namib Desert. Only very recently, Hinchliffe et al. (2017), which used high resolution drone-captured photographic images (using RGB and hyperspectral cameras) to map epilithic lichens' distribution in the Namib Desert gravel plains, mentioned specifically *astrobiology* as a field of application of the results they have generated. They notably showed that advanced photogrammetry represents a valid methodology to use when attempting to assess microbial colonization using rovers or probes at the surface of extraterrestrial planets such as Mars.

Nevertheless, the Namib Desert offers a wealth of lithobionthic refuge niche communities (Ramond et al., 2018), lichen fields (Hinchliffe et al., 2017) and hypersaline environments (Johnson et al., 2017) which are commonly used as terrestrial analogues in astrobiology studies (Sancho et al., 2008; Stivaletta and Barbieri, 2009; Parro et al., 2011). Furthermore, anhydrobiotic cyanobacteria of the genus *Chroococcidiopsis* have been detected in Namib Desert hypolithic and edaphic communities (Makhalanyane et al., 2013). Due to their phototrophic and multi-stress abiotic resistance capabilities (particularly against desiccation, extreme temperatures and radiation) as well as their pioneering colonization of cryptic niches, isolates from this genus have been identified as suitable astrobiology experimental models and selected for space missions (Billi et al., 2013). The central Namib Desert gravel plain soils are also characterized by being Africa's largest gypsum ($CaSO_4 \cdot 2H_2O$) deposit (Eckardt and Spiro, 1999), a sulfate mineral that has been detected as a major constituent of the hydrated region in the north polar regions of Mars (Langevin et al., 2005). Consequently, this Mars analogue substratum represents an ideal system to study potentially relevant Martian-like microbial physiologies. Altogether, these observations clearly show that the Namib Desert sh/could become a new Eldorado for astrobiology and exobiology studies.

8 Conclusions

Since the founding of the Gobabeb Research and Training Centre in 1962, Namib Desert ecological research has expanded to make this one of the most studied hot deserts in the world, along with the Negev and Atacama deserts. The Centre is strategically situated on the banks of the Kuiseb River, which separates the northern gravel plains and southern Namib Sand-Sea. The near-unique climatic regime of the Namib (i.e., a longitudinal water regime transition from coastal fog to inland rain; Eckardt et al., 2013a) has been central to much of the research effort, which has focused on the role of water regimes in speciation and adaptation. The Namib Desert has also proved to be a fascinating environment for studies of the adaptation of microbial communities to poly-extreme conditions—particularly hyperaridity. This is enhanced by the fact that the Namib Desert presents a multitude of landscapes, geologies, lithologies, and cryptic microbial-colonized niches (Seely and Pallett, 2008).

The major focus of Namib Desert microbial ecology research has been on the bacterial, fungal, and to a lesser extent, archaeal communities (e.g., Johnson et al., 2017). Very little attention has been paid, to date, to the viruses (e.g., Adriaenssens et al., 2016) and microeukaryotes (e.g., André et al., 1997), both of which are likely to play critically important roles in the functioning (and maybe control) of desert soil microbial communities.

Culture-independent "omics" technologies have hugely expanded our understanding of the diversity of Bacteria, Archaea, and fungi in Namib Desert soils. Metagenomics, with molecular fingerprinting (e.g., Gombeer et al., 2015), barcoding (e.g., Van der Walt et al., 2016), and shotgun (e.g., Vikram et al., 2016) sequencing [which also encompass meta-viromics studies; e.g., Adriaenssens et al., 2016] studies, have now been applied extensively, while meta-proteomics (Gunnigle et al., 2014) and meta-transcriptomics (Gunnigle et al., 2017) approaches have, as yet, been used only to a limited extent. Metabolomics, a technique which has considerable potential for studies of adaptation (e.g., solute accumulation stress responses) and community carbon flow, has yet to be applied to Namib Desert microbial ecology.

Very little research on functional processes in Namib Desert microbial communities has yet been reported (but see, e.g., Gunnigle et al., 2014, 2017; Ramond et al., 2018). However, given that desert soil and lithic microbial communities are responsible for most ecosystem services for much of any annual cycle (in the absence of higher plants), and given that rising global temperatures, linked to Global Climate Change, are predicted by current climate models to increase the scale of drylands (IPCC, 2014), we suggest that future climate models should include microbial process rates. We suggest that further efforts to unravel microbial process rates in desert ecosystems, together with in situ and in silico experiments mimicking predicted climate change scenarios (e.g., Escolar et al., 2012) would be of considerable value in future climate forecasting.

References

Adriaenssens, E.M., van Zyl, L., de Maayer, P., Rubagotti, E., Rybicki, E., Tuffin, M., Cowan, D.A., 2015. Metagenomic analysis of the viral community in Namib Desert hypoliths. Environ. Microbiol. 17, 480–495.

Adriaenssens, E.M., van Zyl, L.J., Cowan, D.A., Trindade, M.I., 2016. Metaviromics of Namib desert salt pans: a novel lineage of haloarchaeal salterproviruses and a rich source of ssDNA viruses. Viruses 8, 14.

Albrecht, C.F., Joubert, J.J., De Rycke, P.H., 2001. Origin of the enigmatic, circular, barren patches ('Fairy Rings') of the pro-Namib. S. Afr. J. Sci. 97, 23–27.

André, H.M., Noti, M.I., Jacobson, K.M., 1997. The soil microarthropoda of the Namib Desert: a patchy mosaic. J. Afr. Zool. 111, 499–518.

Armstrong, A., Valverde, A., Ramond, J.-B., Makhalanyane, T.P., Jansson, J.K., Hopkins, D.W., Aspray, T.J., Seely, M., Trindade, M.I., Cowan, D.A., 2016. Temporal dynamics of hot desert microbial communities reveal structural and functional responses to water input. Sci. Rep. 6, 34434.

Bates, S.T., Berg-Lyons, D., Caporaso, J.G., Walters, W.A., Knight, R., Fierer, N., 2011. Examining the global distribution of dominant archaeal populations in soil. ISME J. 5, 908–917.

Becker, T., Getzin, S., 2000. The fairy circles of Kaokoland (North-West Namibia)—origin, distribution and characteristics. Basic Appl. Ecol. 159, 149–159.

Belnap, J., 2003. The world at your feet: desert biological soil crusts. Front. Ecol. Environ. 1, 181–189.

Belnap, J., Welter, J.R., Grimm, N.B., Barger, N., Ludwig, J.A., 2005. Linkages between microbial and hydrologic processes in arid and semiarid watersheds. Ecology 86, 298–307.

Billi, D., Baqué, M., Smith, H., McKay, C., 2013. Cyanobacteria from extreme deserts to space. Adv. Microbiol. 3, 80–86.

Brum, J.-R., Sullivan, M.-B., 2015. Rising to the challenge: accelerated pace of discovery transforms marine virology. Nat. Rev. Microbiol. 13, 147–159.

Büdel, B., 2011. Lichens of the Namib Desert—a guide to their identification. Herz 24, 167.

Büdel, B., Wessels, D.C.J., 1986. Parmelia hueana Gyeln., a vagrant lichen from the Namib Desert, SWA/Namibia. I. Anatomical and reproductive adaptations. Dinteria 18, 3–36.

Carreira, C., Staal, M., Falkoski, D., Vries, R.P., Middelboe, M., Brussaard, C.P., 2015. Disruption of photoautotrophic intertidal mats by filamentous fungi. Environ. Microbiol. 17, 2910–2921.

Cavicchioli, R., 2002. Extremophiles and the search for extraterrestrial life. Astrobiology 2, 281–292.

Chan, Y., Lacap, D.C., Lau, M.C.Y., Ha, K.Y., Warren-Rhodes, K.A., Cockell, C.S., Cowan, D.A., McKay, C.P., Pointing, S.B., 2012. Hypolithic microbial communities: between a rock and a hard place. Environ. Microbiol. 14, 2272–2282.

Cooper-Driver, G.A., Wagner, C., Kolberg, H., 2000. Patterns of *Aspergillus niger* var. *phoenicis* (corda) Al-Musallam infection in Namibian populations of *Welwitschia mirabilis* Hook. f. J. Arid Environ. 46, 181–198.

Cox, G.W., 1987. The origin of vegetation circles on stony soils of the Namib Desert near Gobabeb, South West Africa/Namibia. J. Arid Environ. 13, 237–243.

Docherty, K.M., Gutknecht, J.L., 2012. The role of environmental microorganisms in ecosystem responses to global change: current state of research and future outlooks. Biogeochemistry 109, 1–6.

Ebner, M., Miranda, T., Roth-Nebelsick, A., 2011. Efficient fog harvesting by *Stipagrostis sabulicola* (Namib Dune Bushman grass). J. Arid Environ. 75, 524–531.

Eckardt, F.D., Spiro, B., 1999. The origin of sulphur in gypsum and dissolved sulphate in the central Namib Desert, Namibia. Sediment. Geol. 123, 255–273.

Eckardt, F.D., Soderberg, K., Coop, L.J., Muller, A.A., Vickery, K.J., Grandin, R.D., Jack, C., Kapalanga, T., Henschel, J., 2013a. The nature of moisture at Gobabeb, in the Central Namib Desert. J. Arid Environ. 93, 7–19.

Eckardt, F.D., Livingstone, I., Seely, M., Van Holdt, J., 2013b. The surface geology and geomorphology around Gobabeb, Namib Desert, Namibia. Geogr. Ann.: Ser. A Phys. Geogr. 95, 271–284.

Eicker, A., Theron, G.K., Grobbelaar, N., 1982. 'n Mikrobiologiese studie van "kaal kolle"in die Giribesvlakte van Kaokoland, SWA-Namibië. S. Afr. J. Bot. 1, 69–74.

Escolar, C., Martínez, I., Bowker, M.A., Maestre, F.T., 2012. Warming reduces the growth and diversity of biological soil crusts in a semi-arid environment: implications for ecosystem structure and functioning. Philos. Trans. R. Soc. B: Biol. Sci. 367, 3087–3099.

Fairén, A.G., Davila, A.F., Lim, D., Bramall, N., Bonaccorsi, R., Zavaleta, J., et al., 2010. Astrobiology through the ages of Mars: the study of terrestrial analogues to understand the habitability of Mars. Astrobiology 10, 821–843.

Falloon, R.E., 1976. *Curvularia trifolii* as a high-temperature turfgrass pathogen. N. Z. J. Agric. Res. 19, 243–248.

Fidanza, M., Settle, D., Wetzel, H., 2016. An index for evaluating Fairy Ring symptoms in Turfgrass. Hortic. Sci. 51, 1194–1196.

Fierer, N., Leff, J.W., Adams, B.J., Nielsen, U.N., Bates, S.T., Lauber, C.L., Owens, S., Gilbert, J.A., Wall, D.H., Caporaso, J.G., 2012. Cross-biome metagenomic analyses of soil microbial communities and their functional attributes. Proc. Natl. Acad. Sci. U. S. A. 109, 21390–21395.

Foissner, W., Agatha, S., Berger, H., 2002. Soil ciliates (Protozoa, Ciliophora) from Namibia (Southwest Africa), with emphasis on two contrasting environments, the Etosha region and the Namib Desert. Biologiezentrum der Oberösterreichischen Landesmuseums 5, 14–59.

Frossard, A., Gerull, L., Mutz, M., Gessner, M.O., 2012. Disconnect of microbial structure and function: enzyme activities and bacterial communities in nascent stream corridors. ISME J. 6, 680–691.

Frossard, A., Ramond, J.-B., Seely, M., Cowan, D.A., 2015. Water regime history drives responses of soil Namib Desert microbial communities to wetting events. Sci. Rep. 5, 12263.

Gamble, F.M., 1980. Rainfall in the Namib Desert Park. Modoqua 12, 175–180.

Getzin, S., Yizhaq, H., Bell, B., et al., 2016. Discovery of fairy circles in Australia supports self-organization theory. Proc. Natl. Acad. Sci. U. S. A. 113, 3551–3556.

Gilichinsky, D.A., Wilson, G.S., Friedmann, E.I., McKay, C.P., Sletten, R.S., Rivkina, E.M., et al., 2007. Microbial populations in Antarctic permafrost: biodiversity, state, age, and implication for astrobiology. Astrobiology 7, 275–311.

Gombeer, S., Ramond, J.-B., Eckardt, F.D., Seely, M., Cowan, D.A., 2015. The influence of surface soil physicochemistry on the edaphic bacterial communities in contrasting terrain types of the Central Namib Desert. Geobiology 13, 494–505.

Gunnigle, E., Ramond, J.B., Frossard, A., Seely, M., Cowan, D.A., 2014. A sequential co-extraction method for DNA, RNA and protein recovery from soil for future system-based approaches. J. Microbiol. Methods 103, 118–123.

Gunnigle, E., Frossard, A., Ramond, J.-B., Guerrero, L., Seely, M., Cowan, D.A., 2017. Diel-scale temporal dynamics recorded for bacterial groups in Namib Desert soil. Sci. Rep. 7, 40189.

Hedayati, M.T., Pasqualotto, A.C., Warn, P.A., et al., 2007. *Aspergillus flavus*: human pathogen, allergen and mycotoxin producer. Microbiology 153, 1677–1692.

Henschel, J.R., Burke, A., Seely, M., 2005. Temporal and spatial variability of grass productivity in the central Namib Desert. Afr. Study Monogr. S30, 43–56.

Hesse, U., van Heusden, P., Kirby, B.M., Olonade, I., van Zyl, L.J., Trindade, M., 2017. Virome assembly and annotation: a surprise in the Namib Desert. Front. Microbiol. 8, 13.

Hinchliffe, G., Bollard-Breen, B., Cowan, D.A., Doshi, A., Gillman, L.N., Maggs-Kolling, G., et al., 2017. Advanced photogrammetry to assess lichen colonization in the hyper-arid Namib Desert. Front. Microbiol. 8, 2083.

IPCC, 2014. Climate change 2014: synthesis report. In: Pachauri, R.K., Meyer, L.A. (Eds.), Contribution of Working Groups I, II and III to the Fifth Assessment Report of the Intergovernmental Panel on Climate Change. IPCC, Geneva, Switzerland. Core Writing Team. 151 pp.

Irish, J., 1994. The biomes of Namibia, as determined by objective characterisation. Res. Natl. Mus. Bloemfontein 10, 549–592.

Jacobson, K.M., 1997. Moisture and substrate stability determine VA-mycorrhizal fungal community distribution and structure in an arid grassland. J. Arid Environ. 35, 59–75.

Jacobson, K.M., Jacobson, P.J., 1998. Rainfall regulates decomposition of buried cellulose in the Namib Desert. J. Arid Environ. 38, 571–583.

Jacobson, K.M., Jacobson, P.J., Miller Jr., O.K., 2003. The mycorrhizal status of *Welwitschia mirabilis*. Mycorrhiza 3, 13–17.

Jacobson, K., van Diepeningen, A., Evans, S., et al., 2015. Non-rainfall moisture activates fungal decomposition of surface litter in the Namib Sand Sea. PLoS ONE 10. e0126977.

Johnson, R.M., Ramond, J.-B., Gunnigle, E., Seely, M., Cowan, D.A., 2017. Namib Desert edaphic bacterial, fungal and archaeal communities assemble through deterministic processes but are influenced by different abiotic parameters. Extremophiles 21, 381–392.

Juergens, N., 2013. The biological underpinnings of Namib Desert fairy circles. Science 339, 1618–1621.

Juergens, N., Oldeland, J., Hachfeld, B., Erb, E., Schultz, C., 2013. Ecology and spatial patterns of large-scale vegetation units within the central Namib Desert. J. Arid Environ. 93, 59–79.

Kimura, M., Jia, Z.-J., Nakayama, N., Asakawa, S., 2008. Ecology of viruses in soils: past, present and future perspectives. Soil Sci. Plant Nutr. 54, 1–32.

Laity, J.J., 2009. Deserts and Desert Environments. John Wiley & Sons, UK. 360 pp.

Lalley, J.S., Viles, H.A., Copeman, N., Cowley, C., Wiser, S., 2006. The influence of multi-scale environmental variables on the distribution of terricolous lichens in a fog desert. J. Veg. Sci. 17, 831–838.

Lange, O.L., Meyer, A., Zellner, H., Heber, U., 1994. Photosynthesis and water relations of lichen soil crusts: field measurements in the coastal fog zone of the Namib Desert. Funct. Ecol. 8, 253–264.

Lange, O.L., Green, T.G.A., Melzer, B., Meyer, A., Zellner, H., 2006. Water relations and CO_2 exchange of the terrestrial lichen *Teloschistes capensis* in the Namib fog desert:

measurements during two seasons in the field and under controlled conditions. Flora-Morphol. Distrib. Funct. Ecol. Plants 201, 268–280.

Langevin, Y., Poulet, F., Bibring, J.P., Gondet, B., 2005. Sulfates in the north polar region of Mars detected by OMEGA/Mars express. Science 307, 1584–1586.

Le Roux, G.J., 1970. The Microbiology of Sand-Dune Eco-Systems in the Namib Desert, Pp. Stellenbosch University, Stellenbosch.

Le, P.T., Makhalanyane, T.P., Guerrero, L.D., Vikram, S., Van de Peer, Y., Cowan, D.A., 2016. Comparative metagenomic analysis reveals mechanisms for stress response in hypoliths from extreme hyperarid desert. Genome Biol. Evol. 8, 2738–2747.

Lebre, P., de Maayer, P., Cowan, D.A., 2017. Xerotolerant bacteria: surviving through a dry spell. Nat. Rev. Microbiol. 15, 285–296.

Logan, R.F., 1960. The central Namib Desert. South West Africa. National Academy of Sciences. Nat. Res. Council (Washington) 758, 1–162.

Loris, K., Schieferstein, B., 1992. Ecological investigations on lichen fields of the Central Namibia. Vegetation 98, 113–128.

Majid, S.A., Graw, M.F., Chatziefthimiou, A.D., et al., 2016. Microbial characterization of Qatari Barchan sand dunes. PLoS ONE 11. e0161836.

Makhalanyane, T.P., Valverde, A., Lacap, D.C., Pointing, S.B., Tuffin, M.I., Cowan, D.A., 2013. Evidence of species recruitment and development of hot desert hypolithic communities. Environ. Microbiol. Rep. 5, 219–224.

Makhalanyane, T.P., Valverde, A., Gunningle, E., Frossard, A., Ramond, J.-B., Cowan, D.A., 2015. Microbial ecology of hot desert edaphic systems. FEMS Microbiol. Rev. 39, 203–221.

Marasco, R., Mosqueira, M.J., Fusi, M., Ramond, J.-B., Merlino, G., Booth, J.M., et al., 2018. Rhizosheath microbial community assembly of sympatric desert speargrasses is independent of the plant host. Microbiome 6, 215.

Marsh, B., 1987. Micro-arthropods associated with *Welwitschia mirabilis* in the Namib Desert. S. Afr. J. Zool. 22, 89–96.

Mendelsohn, J., Jarvis, A., Roberts, C., Robertson, T., 2002. Atlas of Namibia. A Portrait of the Land and its People. New Africa Books, Kenilworth. 200 pp.

Miller, R.M., 2008. The Geology of Namibia. Vol. 3 Geological Survey of Namibia, Windhoek. 235 pp.

Mucina, L., Rutherford, M.C. (Eds.), 2006. The Vegetation of South Africa, Lesotho and Swaziland. Strelitzia. 19 pp.

Naudé, Y., van Rooyen, M.W., Rohwer, E.R., 2011. Evidence for a geochemical origin of the mysterious circles in the Pro-Namib desert. J. Arid Environ. 75, 446–456.

Nicholson, S.E., 2011. Dryland Climatology. Cambridge University Press, Cambridge. 528 pp.

Noy-Meir, I., 1973. Desert ecosystems: environment and producers. Annu. Rev. Ecol. Syst. 4, 25–51.

Parro, V., de Diego-Castilla, G., Moreno-Paz, M., Blanco, Y., Cruz-Gil, P., Rodríguez-Manfredi, J.A., et al., 2011. A microbial oasis in the hypersaline Atacama subsurface discovered by a life detector chip: implications for the search for life on Mars. Astrobiology 11, 969–996.

Pointing, S.B., 2016. Hypolithic communities. In: Biological Soil Crusts: An Organizing Principle in Drylands. Springer International Publishing, pp. 199–213.

Pointing, S.B., Belnap, J., 2012. Microbial colonization and controls in dryland systems. Nat. Rev. Microbiol. 10, 551–562.

Pointing, S.B., Belnap, J., 2014. Disturbance to desert soil ecosystems contributes to dust-mediated impacts at regional scales. Biodivers. Conserv. 23, 1659–1667.

Prestel, E., Salamitou, S., DuBow, M.S., 2008. An examination of the bacteriophages and bacteria of the Namib desert. J. Microbiol. (Korea) 46, 364–372.

Rajeev, L., da Rocha, U.N., Klitgord, N., et al., 2013. Dynamic cyanobacterial response to hydration and dehydration in a desert biological soil crust. ISME J. 7, 2178–2191.

Ramond, J.-B., Pienaar, A., Armstrong, A., Seely, M., Cowan, D.A., 2014. Niche-partitioning of edaphic microbial communities in the Namib Desert gravel plain Fairy Circles. PLoS ONE 9. e109539.

Ramond, J.-B., Woodborne, S., Hall, G., Seely, M., Cowan, D.A., 2018. Namib Desert primary productivity is driven by cryptic microbial community N-fixation. Sci. Rep. 8, 6921.

Ranawat, P., Rawat, S., 2017. Stress response physiology of thermophiles. Arch. Microbiol. 199, 319–414.

Rietkerk, M., Dekker, S.C., de Ruiter, P.C., et al., 2004. Self-organized patchiness and catastrophic shifts in ecosystems. Science 305, 1926–1929.

Rohwer, F., Thurber, R.-V., 2009. Viruses manipulate the marine environment. Nature 459, 207–212.

Ronca, S., Ramond, J.-B., Jones, B.E., Seely, M., Cowan, D.A., 2015. Namib Desert dune/interdune transects exhibit habitat-specific edaphic bacterial communities. Front. Microbiol. 6, 845.

Sancho, L.G., de la Torre, R., Pintado, A., 2008. Lichens, new and promising material from experiments in astrobiology. Fungal Biol. Rev. 22, 103–109.

Schieferstein, B., Loris, K., 1992. Ecological investigations on lichen fields of the Central Namib. Plant Ecol. 98, 113–128.

Schlesinger, W.H., Reynolds, J.F., Cunningham, G.L., et al., 1990. Biological feedbacks in global desertification. Science 247, 1043–1048.

Scholtz, H., 1972. The soils of the Central Namib Desert with special consideration of the soils in the vicinity of Gobabeb. Madoqua II 1, 33–51.

Scola, V., Ramond, J.-B., Frossard, A., Zablocki, O., Adriaenssens, E.M., Johnson, R.M., Seely, M., Cowan, D.A., 2018. Namib desert soil microbial community diversity, assembly and function along a natural xeric stress gradient. Microb. Ecol. 75, 193–203.

Seely, M.K., 1990. Namib ecology: 25 years of Namib research. Transvaal Mus. Monog. 7, 1–223.

Seely, M., Pallett, J., 2008. Namib: Secrets of a Desert Uncovered. Venture Publications, Windhoek, Namibia 197. pp.

Ségalen, L., Renard, M., Lee-Thorp, J.A., Emmanuel, L., Le Callonnec, L., De Reffélis, M., Senut, B., Pickford, M., Melice, J.-L., 2006. Neogene climate change and emergence of C4 grasses in the Namib, South-Western Africa, as reflected in ratite 13C and 18O. Earth Planet. Sci. Lett. 244, 725–734.

Senut, B., Pickford, M., Ségalen, L., 2009. Neogene desertification of Africa. Compt. Rendus Geosci. 341, 591–602.

Siesser, W.G., 1980. Late Miocene origin of the Benguela upwelling system of northern Namibia. Science 208, 283–285.

Srinivasiah, S., Bhavsar, J., Thapar, K., Liles, M., Schoenfeld, T., Wommack, K.-E., 2008. Phages across the biosphere: contrasts of viruses in soil and aquatic environments. Res. Microbiol. 159, 349–357.

Stivaletta, N., Barbieri, R., 2009. Endoliths in terrestrial arid environments: implications for astrobiology. In: From Fossils to Astrobiology. Springer, Dordrecht, pp. 319–333.

Stomeo, F., Valverde, A., Pointing, S.B., McKay, C.P., Warren-Rhodes, K.A., Tuffin, M.I., Seely, M., Cowan, D.A., 2013. Hypolithic and soil microbial community assembly along an aridity gradient in the Namib Desert. Extremophiles 17, 329–337.

Stone, A.E.C., Thomas, D.S.G., 2013. Casting new light on late quaternary environmental and palaeohydrological change in the Namib Desert: a review of the application of optically stimulated luminescence in the region. J. Arid Environ. 93, 40–58.

Stutz, J.C., Copeman, R., Martin, C.A., Morton, J.B., 2000. Patterns of species composition and distribution of arbuscular mycorrhizal fungi in arid regions of southwestern North America and Namibia, Africa. Can. J. Bot. 78, 237–245.

Tarnita, C.E., Bonachela, J.A., Sheffer, E., et al., 2017. A theoretical foundation for multiscale regular vegetation patterns. Nature 541, 398–401.

Terashima, Y., Fukiharu, T., Fujiie, A., 2004. Morphology and comparative ecology of the fairy ring fungi, *Vascellum curtisii* and *Bovista dermoxantha*, on turf of bentgrass, bluegrass and Zoysiagrass. Mycoscience 45, 251–260.

Theron, G.K., 1979. Die verskynsel van kaal kolle in Kaokoland, Suidwes-Afrika. J. S. Afr. Biol. Soc. 20, 43–53.

Tinley, K.L., 1971. Etosha and the Kaokoveld. Afr. Wild Life 25, 1–16.

Tschinkel, W.R., 2012. The life cycle and life span of Namibian fairy circles. PLoS ONE 7. e38056.

Uhlmann, E., Görke, C., Petersen, A., Oberwinkler, F., 2006. Arbuscular mycorrhizae from arid parts of Namibia. J. Arid Environ. 64, 221–237.

Valverde, A., Makhalanyane, T.P., Seely, M., Cowan, D.A., 2015. Cyanobacteria drive community composition and functionality in rock–soil interface communities. Mol. Ecol. 24, 812–821.

Valverde, A., De Maayer, P., Oberholster, T., Henschel, J., Louw, M.K., Cowan, D.A., 2016. Specific microbial communities associate with the rhizosphere of *Welwitschia mirabilis*, a living fossil. PLoS ONE 11. e0153353.

Van der Walt, A.J., Johnson, R.M., Cowan, D.A., Seely, M., Ramond, J.-B., 2016. Unique microbial phylotypes in Namib Desert dune and gravel plain Fairy Circle soils. Appl. Environ. Microbiol. 82, 4592–4601.

Van Heerden, P.D.R., Swanepoel, J.W., Krüger, G.H.J., 2007. Modulation of photosynthesis by drought in two desert scrub species exhibiting C_3-mode assimilation. Environ. Exp. Bot. 61, 124–136.

Van Rooyen, M.W., Theron, G.K., Van Rooyen, N., et al., 2004. Mysterious circles in the Namib Desert: review of hypotheses on their origin. J. Arid Environ. 57, 467–485.

Velázquez, D., López-Bueno, A., de Cárcer, D.A., et al., 2016. Ecosystem function decays by fungal outbreaks in Antarctic microbial mats. Sci. Rep. 6, 22954.

Vikram, S., Guerrero, L.D., Makhalanyane, T.P., Le, P.T., Seely, M., Cowan, D.A., 2016. Metagenomic analysis provides insights into functional capacity in a hyperarid desert soil niche community. Environ. Microbiol. 18, 1875–1888.

Viles, H.A., Goudie, A.S., 2013. Weathering in the Central Namib, Namibia: controls, processes and implications. J. Arid Environ. 93, 20–29.

Walker, J.J., Pace, N.R., 2007. Endolithic microbial ecosystems. Ann. Rev. Microbiol. 61, 331–347.

Walsh, B., Ikeda, S.S., Boland, G.J., 1999. Biology and management of dollar spot (*Sclerotinia homoeocarpa*); an important disease of turfgrass. Hortscience 34, 13–21.

Ward, J.D., 1987. The Cenozoic succession in the Kuiseb Valley, central Namib Desert. In: Geological Survey of South West Africa/Namibia Memoir.vol. 9. 124 pp.

Warren-Rhodes, K.A., McKay, C.P., Boyle, L.N., et al., 2013. Physical ecology of hypolithic communities in the central Namib Desert: the role of fog, rain, rock habitat and light. J. Geophys. Res. 118, 1451–1460.

Wessels, D., 1989. Lichens of the Namib Desert, South West Africa/Namibia. Dinteria 20, 3–22.

Wessels, D.C.J., van Vuuren, D.R.J., 1986. Landsat imagery—its possible use in mapping the distribution of major lichen communities in the Namib Desert, South West Africa. Modoqua 14, 369.

Wickner, R., 2012. Extrachromosomal Elements in Lower Eukaryotes. Springer-Verlag, USA. 568 pp.

Yaakop, A.S., Chan, K.-G., Ee, R., Lim, Y.L., Lee, S.-W., Mana, F.A., Goh, K.M., 2016. Characteriztion of the mechanism of prolonged adaptation to osmotic stress of *Jeotgalibacillus malaysiensis* via genome and transcriptome sequencing analysis. Sci. Rep. 6, 33660.

Yeaton, R.I., 1988. Structure and function of the Namib dune grasslands: characteristics of the environmental gradients and species distributions. J. Ecol. 76, 744–758.

Zablocki, O., Adriaenssens, E.M., Cowan, D.A., 2016. Diversity and ecology of viruses in hyperarid desert soils. Appl. Environ. Microbiol. 82, 770–777.

Zablocki, O., Adriaenssens, E.M., Frossard, A., Seely, M., Ramond, J.-B., Cowan, D.A., 2017. Metaviromes of extracellular soil viruses along a Namib Desert aridity gradient. Genome Announc. 5. e01470-16.

Zhang, Y., Crous, P.W., Schoch, C.L., Hyde, K.D., et al., 2012. Pleosporales. Fungal Divers. 53, 1–221.

Further reading

Büdel, B., Darienko, T., Deutschewitz, K., Dojani, S., Friedl, T., et al., 2009. Southern African biological soil crusts are ubiquitous and highly diverse in drylands, being restricted by rainfall frequency. Microb. Ecol. 57, 229–247.

Moll, E.J., 1994. The origin and distribution of fairy rings in Namibia. In: Proceedings of the 13th Plenary Meeting AETFAT, Malawi, pp. 1203–1209.

Endolithic microbial communities as model systems for ecology and astrobiology

7

Victoria Meslier, Jocelyne DiRuggiero

Department of Biology, The Johns Hopkins University, Baltimore, MD, United States

Chapter outline

1 Introduction

For 3.4 billion years of its history, microorganisms have inhabited the Earth. They are essential for the evolution of its minerals, its major geochemical cycles, and its atmosphere, yet the extent of their diversity, their metabolic capabilities, and their interactions with the biotic and abiotic components of their ecosystem remain largely unexplored. Addressing the link between microbial activities and the Earth's

Model Ecosystems in Extreme Environments. https://doi.org/10.1016/B978-0-12-812742-1.00007-6

geochemistry and geological record is not only critical to our understanding of the history of life on our planet but it is fundamental to interpret extraterrestrial biosignatures and to guide our search for life elsewhere.

The ongoing extensive exploration of our solar system, and the observation of exoplanets, have the potential to answer a long-standing question in astrobiology, *are we alone?* Current conditions on Mars and technical challenges in studying exoplanets—which will most likely result in the first observable "habitable" planets to be close to the inner edge of the habitable zone (HZ) (Seager, 2013)—force us to consider what the dry limits for life might be for these desert worlds. Hyperarid deserts, such as the Atacama Desert in Northern Chile, are well suited to characterize microbial communities under extreme xeric stress, to determine the dry limits for *life as we know it*, and to provide insights into the habitability of extremely dry extraterrestrial environments.

Soils in hyperarid deserts are colonized by sparse assemblages of specialized microorganisms subjected to extensive water stress and restricted access to organic substrates, resulting in long periods of stasis (Cowan et al., 2002; Navarro-Gonzalez, 2003; Drees et al., 2006; Connon et al., 2007; Crits-Christoph et al., 2013). In contrast, "islands of microbial life" are found inside translucent rocks, in cold and hot deserts around the world, where photosynthetic primary producers support a diversity of heterotrophic bacteria and archaea (Walker and Pace, 2007; Wierzchos et al., 2012a). Under extreme water deficit and high solar radiation fluxes, lithobiontic (within rock) ecological niches are considered environmental refuges for life (Friedmann, 1982; Cary et al., 2010; Pointing and Belnap, 2012). Lithic microbial communities are also good tractable model systems for microbial ecology because they are ubiquitous in deserts around the world and they are low complexity microbial communities (Walker and Pace, 2007).

In this review, we will first discuss the lithic habitat, the various types of rock substrates and their geographic distribution, and the colonization strategies evolved by lithic microorganisms. In the second part, we will examine our state of knowledge on the taxonomy and functional diversity of endolithic microbial communities. We will then focus on the physical and chemical properties of the rock substrates that explain their role as refuges for life in extremely dry deserts and highlight what rock properties are thought to be essential for microbial recruitment. Finally, we will discuss how the study of such communities provides an understanding for the dry limits for life on Earth and how it informs our search for life on other planets and moons.

2 The lithic microbial habitat in hot and cold deserts

2.1 Historical context

The study of lithic (from Greek *lithos*, stone) microbial life in extreme environments was first reported by Friedmann and colleagues in the 1960s, when they demonstrated the presence of algae in sandstones from the Negev Desert in

Israel (Friedmann et al., 1967). Although this was one of the first reports of lithic colonization, it is their following work on endolithic algae from the Dry Valleys of Antarctica that dramatically expended the limits for *life as we know it*, and opened a whole new field of research (Friedmann and Ocampo, 1976). In this region of the world supposedly devoid of life, due to extremely low temperatures and the nearly absence of liquid water, the discovery of such microorganisms has forced the scientific community to reconsider what the dry limits of life are, that is, "the threshold where water is too scare to permit the full range of functions necessary to sustain viable population of microorganisms, and where biological adaptation to desiccation is no longer possible" (Wierzchos et al., 2012a). Since then, extreme habitats on Earth have gained scientific interest and a number of lithic microbial communities have been described from deserts around the world.

2.2 How extreme the driest places on Earth are

Arid environments encompass more than 35% of the world's land area, nearly 60 million square kilometers, and represent the most common habitat on Earth after oceans (Mares, 1999). According to the United Nations Environmental Programme, arid environments can be categorized by their Aridity Index (AI), defined as the ratio between the mean annual precipitation and the mean annual evapotranspiration (Barrow, 1992). Arid deserts (AI < 0.2) experienced low and irregular precipitation events, around 250 mm annually. However, more extreme regions—the so-called hyperarid deserts—experienced an annual mean precipitation below 25 mm (AI < 0.05). A number of hot and cold deserts around the world fits these latei characteristics and are also considered analog environments for Mars (Friedmann, 1986; Domagal-Goldman et al., 2016) (Fig. 1). Under these extreme desert conditions, microorganisms need to cope not only with the scarcity of water but also with high solar fluxes, including lethal UV radiation, high and low temperatures, high rates of evaporation, prolonged periods of desiccation, oligotrophic conditions, and frequently high salinity levels (Wierzchos et al., 2012a).

While hot and cold deserts are considered extremes compared to the rest of the terrestrial biomes, these are not necessarily equals regarding to how extreme they are. For example, Cockell and coworkers (Cockell et al., 2002a) showed that polar deserts, located at almost exact opposite latitudes, differed regarding temperature and water stresses, with the Arctic desert being more clement than the Antarctic desert. Furthermore, Friedmann argued that hot deserts were more extreme than polar deserts, not only regarding air temperature, water, and light availability, but because of the short daily periods when metabolic activity and growth were permitted (Friedmann, 1980).

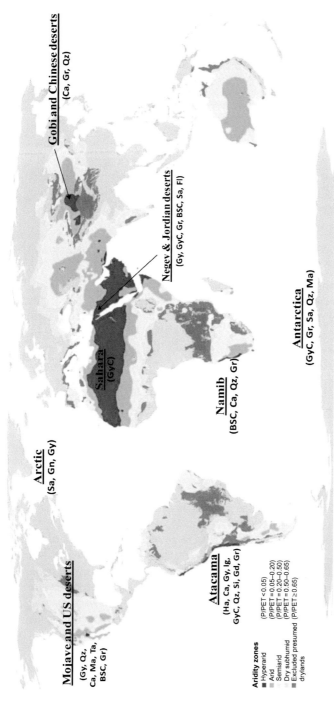

FIG. 1

Hyperarid deserts and rock substrates with known lithic microbial communities around the world. Classification of deserts are according to the United Nations Environment Programme and using the Aridity Index (AI = P/PET). *BSC*, biological soil crust; *Ca*, Carbonates (*Calcite, Limestone, Dolomite*); *Gy*, gypsum; *GyC*, gypsum crust; *Gr*, granite; *Gn*, gneiss; *Gd*, granodiorite; *Ha*, halite; *Ig*, ignimbrite; *Ma*, marble; *Qz*, quartz; *Sa*, sandstone; *Si*, siltstone; *Fl*, flint; *Ta*, talc. Note that Arctic and Antarctic polar deserts were not classified in this map.

Modified from UNEP, http://www.unep.org/.

Note: P = precipitation and PET = potential evapotranspiration.
Source: UNEP-WCMC (2007)

Aridity zones
- Hyperarid (P/PET < 0.05)
- Arid (P/PET = 0.05–0.20)
- Semiarid (P/PET = 0.20–0.50)
- Dry subhumid (P/PET = 0.50–0.65)
- Excluded presumed drylands (P/PET ≥ 0.65)

Mojave and US deserts
(Gy, Qz, Ca, Ma, Ta, BSC, Gr)

Arctic
(Sa, Gn, Gy)

Atacama
(Ha, Ca, Gy, Ig, GyC, Qz, Si, Gd, Gr)

Gobi and Chinese deserts
(Ca, Gr, Qz)

Negev & Jordian deserts
(Gy, GyC, Gr, BSC, Sa, Fl)

Sahara
(GyC)

Namib
(BSC, Ca, Qz, Gr)

Antarctica
(GyC, Gr, Sa, Qz, Ma)

2.3 **The lithic habitat**

In extreme deserts, when environmental conditions become too harsh, microorganisms find refuge on or inside rocks (i.e., lithic habitat) as a stress avoidance strategy (Walker and Pace, 2007). Lithic communities are phototrophic-based with primary producers, typically *Cyanobacteria* and algae, supporting a diversity of consumers, including fungi, and heterotrophic bacteria and archaea (Friedmann, 1982; Omelon, 2008; Nienow, 2009; Wierzchos et al., 2012a). These communities develop at optimal locations underneath the surface of translucent rocks that is called the colonization zone (Friedmann, 1982; Wierzchos et al., 2012a). The overlying mineral substrate provides protection from incident UV radiation and excessive photosynthetically active radiation (PAR), thermal buffering, protection from freeze-thaw events, physical stability, and enhanced water availability (Walker and Pace, 2007). While the susceptibility of a rock to colonization—its bioreceptivity—depends mainly on its physical and chemical properties (Walker and Pace, 2007), properties intrinsic to the microbial community itself are also essential to surviving extreme xeric stress. These can be the production of an extracellular polymeric substance (EPS) matrix with a strong hygroscopic nature (Wierzchos et al., 2012a), the ability to reversibly activate metabolism and grow in the short periods when water is available (Lange et al., 1994), and the capability to delay metabolic activity during dehydration (Harel et al., 2004). Desiccation-tolerant cells implement structural, physiological, and molecular mechanisms to survive severe water deficit and these mechanisms together with their regulation need further understanding (Lebre et al., 2017).

The lithic habitat can be divided into three categories, depending where the microorganisms are found on/or within the substrate (Golubic et al., 1981; Chan et al., 2012; Wierzchos et al., 2012a): (i) in the epilithic habitat, microorganisms colonized the surface of the rock that is directly exposed to the sun; (ii) in the hypolithic habitat, microbial cells are found underneath the rock and are most likely in contact with the soil; and (iii) in the endolithic habitat, microorganisms colonized the interior of the rock, generally occurring within few millimeters under the rock surface. The endolithic habitat can be further divided into three subcategories: (i) the chasmoendolithic habitat, which consists of cracks and fissures directly connected to the rock surface; (ii) the cryptoendolithic habitat, corresponding to pore spaces within the mineral matrix; and (iii) the hypoendolithic habitat, in which microorganisms are found within the underside of the rock. This review will focus more specifically on the endolithic habitat (for papers on the hypolithic habitat, see Warren-Rhodes et al., 2007; Makhalanyane et al., 2013, 2015a; Chan et al., 2012; Pointing and Belnap, 2012).

Over the past decades, direct microscopic visualization of the microbe-rock interface has significantly improved our knowledge on the different modes of endolithic colonization (Wierzchos and Ascaso, 1994). The cryptoendolithic habitat, the most common mode of endolithic colonization, has been described from multiple substrates and deserts, including gypsum (Dong et al., 2007; Wierzchos

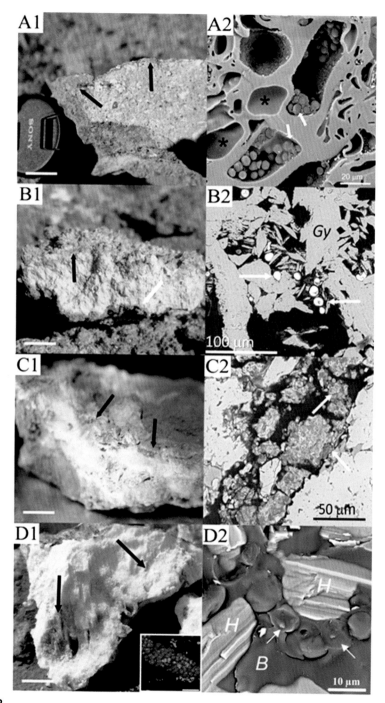

FIG. 2

Endolithic habitats from the Atacama Desert. On the left are cross-sections photographs of various substrates with colonization zone indicated by arrows, and on the right, electron

(Continued)

et al., 2015; Meslier et al., 2018), ignimbrite (Wierzchos et al., 2013; Cámara et al., 2014), halite (Vítek et al., 2012; Robinson et al., 2015), gypsum crust (Hughes and Lawley, 2003; Wierzchos et al., 2011), limestone (McNamara et al., 2006; Matthes et al., 2001), sandstone (Friedmann and Ocampo, 1976), and gneiss (Cockell et al., 2002b) (Fig. 2). In ignimbrite, the cryptoendolithic colonization occurs as a thin layer (up to 2 mm) right under the rock surface and electron microscopy has shown that some of the bottle-shaped pores of the rocks were filled with microorganisms while others, probably not connected to the surface, remained empty (Fig. 2A) (Wierzchos et al., 2013). In the gypsum, cryptoendolithic colonization can be found under the top and bottom surfaces of the rock and as such provides small microhabitats for various microbial assemblages (Fig. 2B) (Wierzchos et al., 2015). In contrast, the calcite chasmoendolithic colonization, in natural perpendicular and parallel cracks and fissures of the rock, provides large spaces for microbial interactions (Fig. 2C) (DiRuggiero et al., 2013). This type of colonization has also been reported for granite rocks from Antarctica (de los Ríos et al., 2005; Archer et al., 2017) and more recently from the Atacama Desert (Meslier et al., 2018). In halite (NaCl rocks) nodules, cyanobacteria colonies can be seen embedded in brine as a result of salt deliquescence, which occurs even at very low air relative humidity (Fig. 2D) (Wierzchos et al., 2006; Robinson et al., 2015).

Early studies of sandstones from the Negev and Antarctica deserts (Friedmann et al., 1967; Friedmann, 1971; Friedmann and Ocampo, 1976), and from the Colorado Plateau (Bell et al., 1986), identified microorganisms by morphology using

---

FIG 2, CONT'D

microscopy photographs of colonization zones with microbial cells indicated by white arrows. (A1) Cryptoendolithic habitat in ignimbrite rock, (A2) LT-SEM image showing pores partially or fully occupied by cells and empty pores (asterisks) in ignimbrite; (B1) cryptoendolithic *(black arrow)* and hypoendolithic *(white arrow)* habitats found in gypsum, (B2) SEM-BSE image showing algal cells and fungal hyphae colonizing pore spaces between gypsum (Gyp) crystals; (C1) chasmoendolithic habitat in calcite showing the colonized fractured substrate, (C2) SEM-BSE image of chasmoendolithic cells in the cracks of the calcite; (D1) cryptoendolithic habitat in halite nodules with in situ CLSM image of SYBR Green stained cells showing cyanobacteria *(red autofluorescence)* and associated heterotrophic bacteria and archaea *(green signal)*; (D2) LT-SEM micrograph of cyanobacterial colonies among halite crystals (asterisks). Scale bars for cross-sections photographs = 1 cm.

A2 from Wierzchos, J., et al., 2013. Ignimbrite as a substrate for endolithic life in the hyper-arid Atacama Desert: implications for the search for life on Mars. Icarus, 224(2), 334–346. B2 from Wierzchos, J., et al., 2015. Adaptation strategies of endolithic chlorophototrophs to survive the hyperarid and extreme solar radiation environment of the Atacama Desert. Front. Microbiol., 6, 1–11. C2 from Crits-Christoph, A., et al., 2016a. Phylogenetic and functional substrate specificity for endolithic microbial communities in hyper-arid environments. Front. Microbiol., 7, 1–15. D1 insert from Robinson, C.K., et al., 2015. Microbial diversity and the presence of algae in halite endolithic communities are correlated to atmospheric moisture in the hyper-arid zone of the Atacama Desert. Environ. Microbiol., 17(2), 299–315. D2 from Robinson, C.K., et al., 2015. Microbial diversity and the presence of algae in halite endolithic communities are correlated to atmospheric moisture in the hyper-arid zone of the Atacama Desert. Environ. Microbiol., 17(2), 299–315.

direct visualization of the cells in their natural habitats. Two community types, solely based on the morphology of the dominant phototrophs were identified; eukaryotic-dominated communities, with mostly the green algae *Trebouxia*, and prokaryotic-dominated communities with cyanobacteria such as *Chroococcidiopsis*, *Leptolyngbya*, and *Gloeocapsa* (Friedmann et al., 1988; de los Ríos et al., 2014). In contrast, because of the difficulty in identifying most bacteria by cell morphology, heterotrophic members of the community were first investigated using culture-based methods (Hirsch et al., 1988; Siebert and Hirsch, 1988). Today, sophisticated photon and electron microscopy methods are widely used to visualize in situ microorganisms in the numerous rocks harboring endolithic life (Fig. 2) (Gleeson et al., 2010; Wierzchos et al., 2011, 2013; Vítek et al., 2013, 2016; Ziolkowski et al., 2013); however, to really comprehend "who is there" additional investigative methods are required.

3 Structure and composition of endolithic microbial communities

Advances in molecular biology have deeply changed our views of endolithic microbial diversity and composition (Walker and Pace, 2007). Notably, molecular methods such as DGGE, TRFLP, PCR-based methods, and more recently next-generation sequencing have brought new insights into community composition and revealed far more diverse and complex communities than previously thought (Fig. 3). While endolithic communities are typically dominated by phototrophs, bacterial phyla including *Actinobacteria*, *Proteobacteria*, *Chloroflexi*, *Deinococcus/Thermus*, *Bacteroidetes*, *Acidobacteria*, *Planctomyces*, *Gemmatimonadetes*, and *Verrucomicrobia*, together with a number of archaeal phyla have also been reported to inhabit the endolithic habitat (de la Torre et al., 2003; Pointing et al., 2009; Gleeson et al., 2010; Li et al., 2013; DiRuggiero et al., 2013; Wang et al., 2013; Robinson et al., 2015; Wierzchos et al., 2015; Chong et al., 2015; Lee et al., 2016; Archer et al., 2017; Meslier et al., 2018).

In contrast, the identification of viruses in endolithic communities is still at an early age but several studies using high-throughput sequencing methods have identified diverse viruses in quartz hypoliths and hypersaline environments (Zablocki et al., 2015; Adriaenssens et al., 2014, 2016; Crits-Christoph et al., 2016b). A remarkable diversity of haloviruses was found in halite nodules from the Atacama Desert and nearly 30 complete viral genomes were reconstructed (Crits-Christoph et al., 2016b). Most interestingly, new groups of viruses were described targeting *Halobacteriales*, the cyanobacterium *Halothece*, and the newly described *Nanohaloarchaeon* (Candidatus *Nanopetramus* SG9). The analysis of the reconstructed genome of *Nanopetramus* revealed the presence of a CRISPR (clustered regularly interspaced short palindromic repeats)/CAS system, known to provide acquired immunity against viruses. This system was directly linked to one of the viral genomes present in the community, suggesting active host-virus interactions within

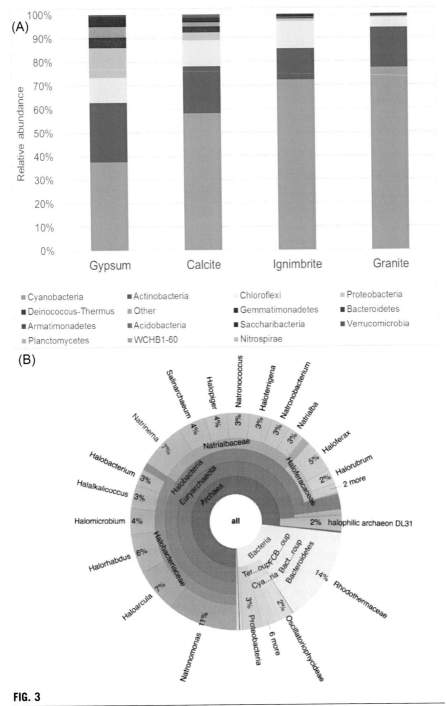

FIG. 3

Taxonomic composition of endolithic microbial communities from the Atacama Desert, Chile.
(A) Relative abundance of phyla in gypsum, calcite, ignimbrite, and granite communities
based on 16S rRNA gene sequences; (B) Krona plot of phylogenetic assignments for
assembled metagenome data of halite (salt) communities. Metagenome sequence reads
were assembled with metaSPAdes and the contigs analyzed with Kraken.

(A) Data from Meslier, V., et al. (2018), Fundamental drivers for endolithic microbial community assemblies in
the hyperarid Atacama Desert. Environ. Microbiol., 20, 1765–1781. (B) Data from Crits-Christoph, A., et al.,
2016a. Phylogenetic and functional substrate specificity for endolithic microbial communities in hyper-arid
environments. Front. Microbiol., 7, 1–15. and analysis from Uritskiy et al. (unpublished)

the salt endolith (Crits-Christoph et al., 2016b). Although the ecological role of viruses in the endolithic habitat is still poorly understood, it is likely that viruses will have an essential role in modulating microbial dynamics in these ecosystems as well (Makhalanyane et al., 2015b).

3.1 The "metacommunity" concept

It is now well established that endolithic microbial communities can include members of the three domains of life and viruses. Similarity at the phylum level between communities from multiple substrates and geographic locations (Makhalanyane et al., 2015b; Lacap-Bugler et al., 2017; Meslier et al., 2018), and the presence of analogous xeric tolerance mechanisms (Potts, 1999; Billi et al., 2000; Kottemann et al., 2005; Bull, 2011; Mohammadipanah and Wink, 2016; Lebre et al., 2017), support the concept of "metacommunity" previously proposed by Walker and Pace (Walker and Pace, 2007). This is the idea that endolithic communities are most likely seeded from a relatively small reservoir of microorganisms that are highly specialized and adapted to the rock environment (Walker and Pace, 2007; Pointing et al., 2009; Chan et al., 2012).

3.2 Highly specialized communities

While the phylogenetic composition of endolithic communities is relatively constrained, significant variations in the structure of these communities were observed between deserts, at the local and global scales, and between substrates (Friedmann, 1980; de la Torre et al., 2003; Pointing et al., 2009, 2015; Chan et al., 2012; Crits-Christoph et al., 2016a; Archer et al., 2017). In the Atacama Desert, where a number of endolithic habitats have been reported, studies showed significant divergence in the relative abundance of *Cyanobacteria*, *Actinobacteria*, *Proteobacteria*, and *Chloroflexi* between gypsum, ignimbrite, granite, and calcite endoliths (Fig. 3A), suggesting substrate-specific adaptations (Dong et al., 2007; DiRuggiero et al., 2013; Wierzchos et al., 2015; Crits-Christoph et al., 2016a; Meslier et al., 2018).

Adaptations to high salt are also essential for communities inhabiting halite nodules in Atacama Desert Salars. These communities are dominated by *Halobacteriales*, *Bacteroidetes*, and *Cyanobacteria*, with, in some instances, the presence of algae (Fig. 3B) (Robinson et al., 2015; Crits-Christoph et al., 2016b; Finstad et al., 2017; Uritskiy et al., 2019). Algae were also found in gypsum where the colonization zone is vertically organized with a layer of scytonemin-productive *Cyanobacteria* beneath a protective layer of carotenoid-containing algae (Wierzchos et al., 2015). Specific attention has been brought to the biogeography of cyanobacteria because of their versatility across hyperarid deserts and significant contribution as primary producers. *Cyanobacteria* phylotypes were found specific to either hot or cold deserts, with a significant difference in the ratio of the two most abundant taxa; *Chroococcidiopsis* was mostly present in hot deserts while *Phormidium* was almost exclusively found in polar deserts (Lacap-Bugler et al., 2017). These authors suggested that endolithic *Cyanobacteria* have not experienced gene flow for extended time scales and may be considered phylogenetically endemic (Bahl et al., 2011; Archer et al., 2017).

3.3 Functional diversity and physiological activity

Metagenome sequencing has given us full access to the genetic complement of whole microbial communities, allowing for a better understanding of community functions, interactions, and adaptations. Functional analysis of calcite and ignimbrite metagenomes from the Atacama Desert revealed a strong variability in community composition and diversity but a surprisingly high similarity at the functional level, with many core functions shared by a large number of species (Fig. 4) (Crits-Christoph

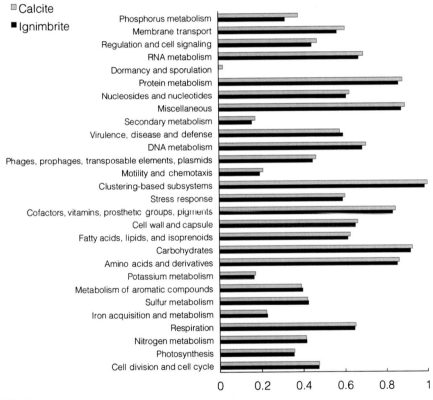

FIG. 4

Functional categories for calcite and ignimbrite endolithic communities from the Atacama Desert. Functional categories were obtained with the SEED subsystems from the respective metagenomes sequences.

From Crits-Christoph, A., et al., 2016a. Phylogenetic and functional substrate specificity for endolithic microbial communities in hyper-arid environments. Front. Microbiol., 7, 1–15.

et al., 2016a). Functional redundancy has also been described for soil and lithic communities from the Antarctic polar desert (Chan et al., 2013; Wei et al., 2016), suggesting that detrimental effects to community functioning and survival would be minimized in case of the loss of an individual species (Chong et al., 2015). However, communities inhabiting different substrates also showed functional differences, indicating environmental selection for specific traits. For example, the metagenome of the Atacama Desert ignimbrite community, when compared to that of the calcite, was enriched in pathways for secondary metabolites, such as nonribosomal peptides (NRPS) and polyketides (PKS) encoding for iron acquisition and antimicrobial compounds. This strongly suggested a harsh competition for space and resources in the narrow pores of the ignimbrite rock in contrast to the large cracks and fissures of the calcite rock (Fig. 2A and C). A notable absence reported from all metagenomes from the Atacama Desert is that of genes for nitrogen fixation and diazotrophy (Crits-Christoph et al., 2016a, 2016b; Finstad et al., 2017). The accumulation of high concentration of nitrate and ammonia is occurring in soils of the Atacama Desert, as a result of atmospheric deposition over millions of years (Michalski et al., 2004), and these compounds are most likely the main source of nitrogen for soil and lithic communities of the desert.

Functional analysis from metagenomics data provides an overview of the metabolic potential of a community but it does answer question: are these communities metabolically active under environmental conditions? Recently, metabolic activity inside halite nodules was demonstrated by in situ measurements of variable chlorophyll fluorescence, using Pulse Amplitude Modulation (PAM) fluorometry, and respiration, using oxygen microprobes (Davila et al., 2015). Controlled laboratory experiments also revealed that the halite community is capable of carbon fixation both through oxygenic photosynthesis and potentially ammonia oxidation (Davila et al., 2015). Although difficult, rigorous field-based investigations, combined with molecular studies and physiological experiments, are critical to further comprehend the functioning of endolithic microbial communities. Another open question regarding the endolithic habitat is that of the biotic and abiotic factors that drive the assembly of these communities.

4 The rock substrate: an essential driver of microbial colonization

Severe stresses are imposed on endolithic communities from hot and cold deserts by the extreme climate conditions but substrate physical and chemical properties, that is, its bioreceptivity, also appear to be essential in the assembly of these communities (Wierzchos et al., 2012a). The substrate's characteristics shape a complex and delicate balance between the amount of water, light, nutrients, and space available to the microorganisms within the rock matrix.

4.1 **Water uptake and retention**

Water is essential for life and for the photosynthetic activity that sustains endolithic communities (Walker and Pace, 2007; Wierzchos et al., 2012a). Liquid water is required for photosynthesis in *Cyanobacteria*; in contrast, algae have been reported to use vapor water above 70% relative humidity for photosynthesis (Lange et al., 1986, 1993; Friedmann et al., 1988; Palmer and Friedmann, 1990; Kidron and Temina, 2017). In hyperarid deserts, precipitations are scarce and endolithic communities often rely on additional sources of water that include fog deposition, dew formation, mineral deliquescence, and snow melted by solar heating in polar deserts (Nienow et al., 1988; Palmer and Friedmann, 1990; Agam and Berliner, 2006). For example, in the calcite chasmoendolithic habitat water penetrates the cracks and fissures connected to the rock surface and the water budget is enhanced by the presence of a crust allowing dew formation on its surface (Budel et al., 2008; Agam and Berliner, 2006; DiRuggiero et al., 2013).

Additionally, the physical and chemical properties of the rock (pores or cracks, their sizes and connection to the rock surface, mineralogy, and chemistry) are critical factors in the adsorption and retention of water. Yet, the determination of water retention capabilities in natural rock substrates, and thus the amount of water available to organisms at the microscale, is very difficult because of the heterogeneity of the substrate itself. This includes (i) irregular cracks and fissures (DiRuggiero et al., 2013; Crits-Christoph et al., 2016a); (ii) various pore sizes not all connected (Wierzchos et al., 2013, 2015); and (iii) heterogeneous mineralogy, as in gypsum, where dispersed sepiolite minerals can absorb and retain great amounts of water (Wierzchos et al., 2015; Meslier et al., 2018). In the cryptoendolithic habitat, water enters the pores by percolation (Friedmann and Ocampo-Friedmann, 1984) and the size of the pores, their homogeneity, and the presence of a crust have all been showed to be essential in the intrarock moisture content (Wierzchos et al., 2013, 2015; Cockell et al., 2002a).

One unique mechanism of water adsorption and retention is that of mineral deliquescence in halite nodules (Davila et al., 2008). The deliquescence of halite (NaCl) occurs when the air relative humidity reaches at least 75%. Under these conditions, water condenses inside the halite pores into concentrated brine, providing liquid water to the community. In the Yungay area of the Atacama Desert, one of the driest places on Earth where no rain, fog, or dew can be detected for years at a time, only the hygroscopic properties of salt provide enough water for colonization of the endolithic habitat (Wierzchos et al., 2012b; Davila et al., 2013) and colonization is virtually absent from all other substrates (Warren-Rhodes et al., 2006; Wierzchos et al., 2011). In addition, it has been shown that the small size of the halite pores, and their distribution inside the nodule, could (i) limit water evaporation, (ii) maintain brine water inside the rock up to 40 days when atmospheric relative humidity (RH) drops <40%, and (iii) retain enough liquid water to activate cyanobacteria photosynthetic activity (Wierzchos et al., 2012b; Davila et al., 2013).

4.2 Light transmission

In addition to liquid water, light is essential for photosynthetic activity. The translucent to semitranslucent properties of a substrate allow light to penetrate the upper layer of the rock while diminishing its intensity (Nienow et al., 1988). In contrast, UV radiations are drastically attenuated and even totally filtered out (Hughes and Lawley, 2003; Villar et al., 2005; Smith et al., 2014; Wierzchos et al., 2015). The mineral composition of a rock substrate, its micromorphology, and its colour are fundamental to its translucence properties, such as, for example, the presence of quartz grains that can increase the amount of light within the colonization zone (Nienow et al., 1988) (Table 1). This shield protection by the substrate to intense solar irradiance is of great importance and it has been shown that endolithic photosynthetic microorganisms are adapted to low light intensities (Bell, 1993; Boison et al., 2004; Dong et al., 2007; Tracy et al., 2010; Raanan et al., 2015). Under such conditions, the minimum light required

Table 1 Light transmittance properties of endolithic habitats

Substrate type	Location	Depth of the endolithic colonization (mm)	% Transmitted light	References
Gypsum crust	Devon Island, Canadian High Arctic	~1	<0.5[a]	Cockell et al. (2010)
	Alexander Island, Antarctica Peninsula	~1 2.5	1 <0.01	Hughes and Lawley (2003)
Gypsum	Cordon de Lila, Atacama	2–5	<1	Wierzchos et al. (2015)
	El Jaroso ravine, Spain[b]	0.5	20	Amaral et al. (2007)
Granite	Zucchelli Station, Antarctica	2	~10	Hall et al. (2008)
Dolomite	Alps, Central Switzerland[b]	1	<5	Horath et al. (2006)
Limestone	Niagara Escarpment, Canada	<5	0.01	Matthes et al. (2001)
Shocked Gneiss	Devon Island, Canadian High Arctic	0.5	<30[a]	Pontefract et al. (2014)

For each substrate we reported the % of transmitted light measured at the depth (mm) of the colonization zone.
[a]Transmission declined with decrease in wavelength.
[b]Not hyperarid deserts.

for photosynthesis was reported to be as low as 0.1 $\mu mol \ m^{-2} \ s^{-1}$ (Raven et al., 2000; Warren-Rhodes et al., 2013).

While some authors found no statistical correlations between transmitted light and depth of the colonization zone (Matthes et al., 2001), numerous other studies have shown that the depth of colonization was constrained partly by the transmitted light (Ferris and Lowson, 1997; Hughes and Lawley, 2003; Horath et al., 2006; Hall et al., 2008; Cockell et al., 2010; DiRuggiero et al., 2013; Wierzchos et al., 2015; Meslier et al., 2018). For example, Horath and coworkers determined the position of the colonization zone in 80 randomly harvested rocks exposed or not to sunlight and found that the colonization zone was significantly deeper and wider in the sunlight-exposed rocks (Horath et al., 2006). Similarly, calcite rocks in the Atacama Desert with wide cracks and fissures, which most likely allow deeper penetration of light, exhibited a wider and deeper colonization zone than other substrates such as ignimbrite and gypsum (DiRuggiero et al., 2013; Crits-Christoph et al., 2016a; Meslier et al., 2018). As such, the size and depth of the colonization is most likely dependent on the intensity and quality of the light transmitted through the substrate.

4.3 Nutrients and microbe-minerals interactions

Endolithic habitats are generally considered oligotrophic environments (Walker and Pace, 2007; Omelon, 2016; Armstrong et al., 2016). In some cases, atmospheric deposition and dust might influence nutrient supplies, such as that of nitrate and ammonia in the Atacama Desert (Michalski et al., 2004; Omelon, 2008; Tang et al., 2012). In situ microbe–mineral interactions are also of importance and while a study showed that the addition of nutrients, such as manganese and phosphate, did not stimulate growth in sandstone endoliths (Johnston and Vestal, 1991), others have shown the release of nutrients by *Cyanobacteria* and *Fungi* by acid production and partial dissolution of the surrounding rock (Garcia-Pichel, 2006; Omelon, 2008; Pontefract et al., 2014). More recently, Jones and Bennett reported that interactions between microorganisms and minerals amended in C and P resulted in a shift in community structure and composition (Jones and Bennett, 2014, 2017), supporting the idea that the mineralogy of the substrate can be a notable determinant in the structure of endolithic communities.

4.4 Architecture and space available

The concept of the rock architecture, that is, the space available for colonization, embodied by the size of the cracks, fissures, and pores and their connection to the surface, has recently been introduced as an important factor, which, linked to light penetration and water retention, is ultimately a driver of community diversity and composition (Meslier et al., 2018; Wierzchos et al., 2015; Crits-Christoph et al., 2016a). For example, in cryptoendolithic habitats, such as ignimbrite or sandstone, the small pores might lead to diffusion limitation of nutrients and metabolic by-products within the community, thus impacting metabolic rates, community functioning, and ultimately the diversity of the microbial assemblages (Crits-Christoph et al., 2016a). In ignimbrite, these diffusional constraints are exacerbated by additional stress to the community from competitions for spare resources and the effects of antimicrobial compounds (Crits-Christoph et al., 2016a). In contrast, the spacious spaces found in gypsum and calcite rocks allow for better diffusion rates and interactions between large aggregates of microorganisms, promoting nutrient exchanges and a diversity of metabolisms (Wierzchos et al., 2015; Crits-Christoph et al., 2016a; Meslier et al., 2018).

5 Implications for astrobiology

Many of the exoplanets discovered so far orbit stars younger than the sun (Fig. 5) and, because for most its history life on Earth was microbial, it is more likely that these planets are inhabited by microbial life than by more advanced life forms. In fact, more than 3891 exoplanets have been discovered and confirmed as of February 2019 and a few thousand more planet "candidates" are waiting to be confirmed (https://exoplanets.nasa.gov/). In 2017, the discovery of seven Earth-size planets orbiting the ultracool dwarf star TRAPPIST-1 (Gillon et al., 2017; Luger et al.,

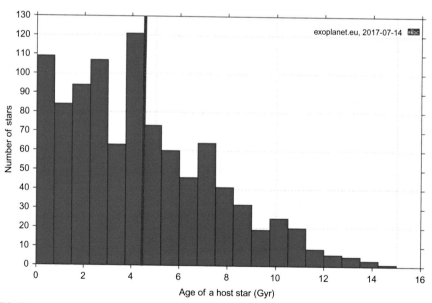

FIG. 5

Number of stars with confirmed planets as a function of the age of the host stars. In red the age of the Sun and the Earth ~4.5 Gyr.

Data from the Exoplanet.eu database.

2017), some 39 light years from Earth, has generated a lot of excitement in the field. In view of such numbers and diversity of planets, a well-constrained understanding of what makes a planet habitable is necessary to maximize our chances to identify a habitable world (Seager, 2013). Currently, the boundaries of what is called the HZ—the region around a star where a planet can have surface liquid water (Lang, 1986)—are subject to intense debates (Kasting et al., 1993; Kopparapu, 2013; Seager, 2013; Zsom et al., 2013). Given the challenges of telescopic observations of exoplanet environments and the fact that planet detection techniques are more sensitive to smaller planets around smaller stars, it is likely that the first potentially habitable terrestrial planets amenable to spectroscopic follow-up will be found orbiting an M-dwarf in our Solar neighborhood (Seager, 2013). These planets will be on close-in orbits around their stars, might be tidally locked, and will have surface temperatures higher than those of Earth—so-called *desert worlds* (Zsom et al., 2013). Therefore being able to constrain the inner edge of the HZ becomes critical since it will most likely be where we will observe and study the next "habitable" planets. However, discrepancies between astronomical HZ models (Kasting et al., 1993; Kopparapu, 2013; Seager, 2013; Zsom et al., 2013) and what we know about life on Earth could extend the inner edge of the HZ to environments that could not be habitable for Life as we know it. Therefore current studies to determine the factors driving the colonization of lithobiontic microbial habitats in the driest deserts on Earth will lead to better constraints for the dry limits for life in our solar system and more accurate models for the inner edge of the HZ.

While a number of studies provided evidence that liquid water existed on the surface of Mars, possibly as recently as 300,000 years ago (e.g., Klingelhoder et al., 2004; Reiss et al., 2004; Squyres et al., 2004; Fassett and Head, 2006; Bishop et al., 2008; Mustard et al., 2008; Niles et al., 2010; McEwen et al., 2011), today the Martian surface is cold and dry and it is likely that any past—and possibly extant—microbial life on Mars would have adapted to extremely desiccating conditions. While the Mars regolith might be uninhabitable, chloride- and sulfate-bearing deposits have been discovered in many areas of the planet (Osterloo et al., 2008) and ignimbrite rocks were tentatively identified in Gale Crater, the landing site of the Mars Science Laboratory (MSL) mission (Wierzchos et al., 2012a). The existence of Martian habitats similar to habitats capable of supporting phototrophic and heterotrophic life in the Atacama Desert—and other hyperarid deserts around the world—suggests that these should be considered important analogs for the search for extant and past life on Mars. Investigating endolithic habitats in the extreme deserts on Earth is therefore a well-suited objective for assessing the survivability of Earth microorganisms under non-Earth conditions (e.g., Mars).

With the potential for a Mars sample return mission and the task of analyzing mineralogical samples for traces of life—so-called biosignatures—it is essential to improve our knowledge of mineral alterations as a result of biological activity and how to distinguish biologically induced structure and organization patterns from inorganic phenomena. "Biosignatures are morphological, chemical (organic, elemental, and/or mineral), and isotopic traces of organisms preserved in minerals, sediments, and rocks" (Westall and Cavalazzi, 2011). They can be the remnants of the organisms themselves, their macromolecules, and/or evidence of their metabolic activities. An exciting possibility is that ancient life forms, or the microstructural or chemical imprint of these life forms, that existed under less restrictive environmental conditions than the current Martian regolith might be conserved within endolithic habitats. It is therefore critical to further our understanding of the interactions between minerals and microorganisms in colonized terrestrial analogs to enhance the possibilities of identifying traces of past life in valuable samples from extraterrestrial environments.

For this, of course, we have to make the assumption that the biochemistry used by terrestrial life is a viable analog for life that might exist elsewhere but this is the only example of life we have so far!

Acknowledgements

This work was founded by grants NNX15AP18G and NNX15AK57G from NASA and grant DEB1556574 from the National Science Foundation to JDR.

References

Adriaenssens, E.M., et al., 2014. Metagenomic analysis of the viral community in Namib Desert hypoliths. Environ. Microbiol. 17 (2), 480–495.

Adriaenssens, E.M., et al., 2016. Metaviromics of Namib desert salt pans: a novel lineage of haloarchaeal salterproviruses and a rich source of ssDNA viruses. Viruses 8 (1), 17–19.

Agam, N., Berliner, P.R., 2006. Dew formation and water vapor adsorption in semi-arid environments—a review. J. Arid Environ. 65 (4), 572–590.

Amaral, G., Martinez-Frias, J., Vazquez, L., 2007. UV shielding properties of Jarosite vs Gypsum: astrobiological implications for Mars. World Appl. Sci. J. 2 (2), 112–116.

Archer, S.D.J., et al., 2017. Endolithic microbial diversity in sandstone and granite from the McMurdo Dry Valleys, Antarctica. Polar Biol. 40, 997–1006.

Armstrong, A., et al., 2016. Temporal dynamics of hot desert microbial communities reveal structural and functional responses to water input. Sci. Rep. 29 (6), 34434.

Bahl, J., et al., 2011. Ancient origins determine global biogeography of hot and cold desert cyanobacteria. Nat. Commun. 25 (2), 163.

Barrow, C.J., 1992. In: Middleton, N., Thomas, D.S.G. (Eds.), World Atlas of Desertification (United Nation Environment Programme). Edward Arnold, London.

Bell, R., 1993. Cryptoendolithic algae of hot semiarid lands and deserts. J. Phycol. 29, 133–139.

Bell, R.A., Athey, P.V., Sommerfeld, M.R., 1986. Cryptoendolithic algal communities of the Colorado plateau. J. Phycol. 22, 429–435.

Billi, D., et al., 2000. Ionizing-radiation resistance in the desiccation-tolerant cyanobacterium Chroococcidiopsis. Appl. Environ. Microbiol. 66 (4), 1489–1492.

Bishop, J., et al., 2008. Phyllosilicate diversity and past aqueous activity revealed at Mawrth Vallis, Mars. Science 321, 830–834.

Boison, G., et al., 2004. Bacterial life and dinitrogen fixation at Gysum rock. Appl. Environ. Microbiol. 70 (12), 7070–7077.

Budel, B., et al., 2008. Dewfall as a water source frequently activates the endolithic cyanobacterial communities in the granites of Taylo Valley, Antarctica. J. Phycol. 44, 1415–1424.

Bull, A.T., 2011. Actinobacteria of the extremobiosphere. In: Horikoshi, K. (Ed.), Extremophiles Handbook. Springer, Japan, pp. 1204–1231.

Cámara, B., et al., 2014. Ignimbrite textural properties as determinants of endolithic colonization patterns from hyper-arid Atacama Desert. Int. Microbiol. 17, 235–247.

Cary, S.C., et al., 2010. On the rocks: the microbiology of Antarctic Dry Valley soils. Nat. Rev. Microbiol. 8 (2), 129–138.

Chan, Y., et al., 2012. Hypolithic microbial communities: between a rock and a hard place. Environ. Microbiol. 14 (9), 2272–2282.

Chan, Y., et al., 2013. Functional ecology of an Antarctic Dry Valley. Proc. Natl. Acad. Sci. U. S. A. 110 (22), 8990–8995.

Chong, C.W., Pearce, D.A., Convey, P., 2015. Emerging spatial patterns in Antarctic prokaryotes. Front. Microbiol. 6, 1–14.

Cockell, C., McKay, C.P., Omelon, C., 2002a. Polar endoliths – an anti-correlation of climatic extremes and microbial biodiversity. Int. J. Astrobiol. 1 (4), 305–310.

Cockell, C.S., et al., 2002b. Impact-induced microbial endolithic habitats. Meteorit. Planet. Sci. 37, 1287–1298.

Cockell, C.S., et al., 2010. The microbe-mineral environment and gypsum neogenesis in a weathered polar evaporite. Geobiology 8 (4), 293–308.

Connon, S.A., et al., 2007. Bacterial diversity in hyperarid Atacama desert soils. J. Geophys. Res. 112 (4), 1–9.

Cowan, D.A., et al., 2002. Antarctic Dry Valley mineral soils contain unexpectedly high levels of microbial biomass. Extremophiles 6 (5), 431–436.

Crits-Christoph, A., et al., 2013. Colonization patterns of soil microbial communities in the Atacama Desert. Microbiome 1, 1–13.

Crits-Christoph, A., et al., 2016a. Phylogenetic and functional substrate specificity for endolithic microbial communities in hyper-arid environments. Front. Microbiol. 7, 1–15.

Crits-Christoph, A., et al., 2016b. Functional analysis of the archaea, bacteria, and viruses from a halite endolithic microbial community. Environ. Microbiol. 18 (6), 2064–2077.

Davila, A.F., et al., 2008. Facilitation of endolithic microbial survival in the hyperarid core of the Atacama Desert by mineral deliquescence. J. Geophys. Res.: Biogeosci. 113(1).

Davila, A.F., et al., 2013. Salt deliquescence drives photosynthesis in the hyperarid Atacama Desert. Environ. Microbiol. Rep. 5 (4), 583–587.

Davila, A.F., et al., 2015. In situ metabolism in halite endolithic microbial communities of the hyperarid Atacama Desert. Front. Microbiol. 6, 1–8.

de la Torre, J., et al., 2003. Microbial diversity of cryptoendolithic communities from the McMurdo Dry Valleys, Antarctica. Appl. Environ. Microbiol. 69 (7), 3858–3867.

de los Ríos, A., et al., 2005. Ecology of endolithic lichens colonizing granite in continental Antarctica. Lichenologist 37 (5), 383.

de los Ríos, A., Wierzchos, J., Ascaso, C., 2014. The lithic microbial ecosystems of Antarctica's McMurdo Dry Valleys. Antarct. Sci. 19 (5), 1–19.

DiRuggiero, J., et al., 2013. Microbial colonisation of chasmoendolithic habitats in the hyperarid zone of the Atacama Desert. Biogeosciences 10 (4), 2439–2450.

Domagal-Goldman, S.D., et al., 2016. The astrobiology primer v2.0. Astrobiology 16 (8), 561–653.

Dong, H., et al., 2007. Endolithic cyanobacteria in soil gypsum: occurrences in Atacama (Chile), Mojave (United States), and Al-Jafr Basin (Jordan) deserts. J. Geophys. Res.: Biogeosci. 112 (2), 1–11.

Drees, K.P., et al., 2006. Bacterial community structure in the hyperarid core of the Atacama Desert, Chile. Appl. Environ. Microbiol. 72 (12), 7902–7908.

Fassett, C.I., Head, J., 2006. Valleys on Hecates Tholus, Mars: origin by basal melting of summit snowpack. Planet. Space Sci. 54, 370–378.

Ferris, F.G., Lowson, E.A., 1997. Ultrastructure and geochemistry of endolithic microorganisms in limestone of the Niagara escarpment. Can. J. Microbiol. 43 (3), 211–219.

Finstad, K.M., et al., 2017. Microbial community structure and the persistence of cyanobacterial populations in salt crusts of the Hyperarid Atacama Desert from genome-resolved metagenomics. Front. Microbiol. 8, 1–10.

Friedmann, I.E., 1971. Light and scanning electron microscopy of the endolithic desert algal habitat. Phycologia 10 (4), 411–428.

Friedmann, E.I., 1980. Endolithic microbial life in hot and cold deserts. Origins of Life 10 (3), 223–235.

Friedmann, E.I., 1982. Endolithic microorganisms in the Antarctic Cold Desert. Adv. Sci. 215 (4536), 1045–1053.

Friedmann, E.I., 1986. The antarctic cold desert and the search for traces of life on Mars. Adv. Space Res. 6 (12), 265–268.

Friedmann, E.I., Ocampo, R., 1976. Endolithic blue-green algae in the dry valleys: primary producers in the Antarctic Desert ecosystem. Science 193 (4259), 1247–1249.

Friedmann, I.E., Ocampo-Friedmann, R., 1984. Endolithic microorganisms in extreme dry environments: analysis of a lithobiontic microbial habitat. In: Klug, M.J., Reddy, C.A. (Eds.), Current Perspectives in Microbial Ecology. ASM Press, pp. 177–185.

Friedmann, I., Lipkin, Y., Ocampo-Paus, R., 1967. Desert algae of the Negev (Israel). Phycologia 6 (4), 185–200.

Friedmann, E.I., Hua, M., Ocampo-Friedmann, R., 1988. Cryptoendolithic lichen and cyanobacterial communities of the Ross Desert, Antarctica. Polarforschung 58 (2–3), 251–259.

Garcia-Pichel, F., 2006. Plausible mechanisms for the boring on carbonates by microbial phototrophs. Sediment. Geol. 185, 205–213.

Gillon, M., et al., 2017. Seven temperate terrestrial planets around the nearby ultracool dwarf star TRAPPIST-1. Nature 542 (7642), 456–460.

Gleeson, D.B., et al., 2010. Molecular characterization of fungal communities in sandstone. Geomicrobiol J. 27 (6–7), 559–571.

Golubic, S., Friedmann, I., Schneider, J., 1981. The lithobiontic ecological niche, with special reference to microorganisms. J. Sediment. Res. 51 (2), 475–478.

Hall, K., Guglielmin, M., Strini, A., 2008. Weathering of granite in Antarctica: I. Light penetration into rock and implications for rock weathering and endolithic communities. Earth Surf. Process. Landf. 33 (2), 295–307.

Harel, Y., Ohad, I., Kaplan, A., 2004. Activation of photosynthesis and resistance to photoinhibition in cyanobacteria within biological desert crust. Plant Physiol. 136 (2), 3070–3079.

Hirsch, B.P., et al., 1988. Diversity and identification of heterotrophs from Antarctic rocks of the McMurdo dry valleys (Ross Desert). Polarforschung 58 (213), 261–269.

Horath, T., Neu, T.R., Bachofen, R., 2006. An endolithic microbial community in dolomite rock in Central Switzerland: characterization by reflection spectroscopy, pigment analyses, scanning electron microscopy, and laser scanning microscopy. Microb. Ecol. 51 (3), 353–364.

Hughes, K.A., Lawley, B., 2003. A novel Antarctic microbial endolithic community within gypsum crusts. Environ. Microbiol. 5 (7), 555–565.

Johnston, C.G., Vestal, J.R., 1991. Photosynthetic carbon incorporation and turnover in Antarctic cryptoendolithic microbial communities: are they the slowest-growing communities on earth? Appl. Environ. Microbiol. 57 (8), 2308–2311.

Jones, A.A., Bennett, P.C., 2014. Mineral microniches control the diversity of subsurface microbial populations. Geomicrobiol J. 31 (3), 246–261.

Jones, A.A., Bennett, P.C., 2017. Mineral ecology: surface specific colonization and geochemical drivers of biofilm accumulation, composition, and phylogeny. Front. Microbiol. 8, 1–14.

Kasting, J.F., Whitmire, D.P., Reynolds, R.T., 1993. Habitable zones around main sequence stars. Icarus 101, 108–128.

Kidron, G.J., Temina, M., 2017. Non-rainfall water input determines lichen and cyanobacteria zonation on limestone bedrock in the Negev highlands. Flora: Morphol. Distrib. Funct. Ecol. Plants 229, 71–79.

Klingelhoder, G., et al., 2004. Jarosite and hematite at Meridiani Planum from opportunity's Mossbauer spectrometer. Science 306, 1740–1745.

Kopparapu, R. 2013. A revised estimate of the occurrence rate of terrestrial planets in the habitable zones around kepler m-dwarfs. *arXiv:1303.2649v1*, pp. 1–12.

Kottemann, M., et al., 2005. Physiological responses of the halophilic archaeon Halobacterium sp. strain NRC1 to desiccation and gamma irradiation. Extremophiles 9 (3), 219–227.

Lacap-Bugler, D.C., et al., 2017. Global diversity of desert hypolithic cyanobacteria. Front. Microbiol. 8, 1–13.

Lang, E., 1986. Physical-chemical limits for the stability of biomolecules. Adv. Space Res. 6, 251–255.

Lange, O.L., Kilian, E., Ziegler, H., 1986. Water vapor uptake and photosynthesis of lichens: with green and blue-green algae as phycobionts. Oecologia 71, 104–110.

Lange, O.L., et al., 1993. Further evidence that activation of photosynthesis by dry cyanobacterial lichens requires liquid water. Lichenologist 25 (2), 175–189.

Lange, O.L., Meyer, A., Budel, B., 1994. Net photosynthesis activation of a desiccated cyanobacterium without liquid water in high air humidity alone—experiments with *Microcoleus sociatus* isolated from a desert soil crust. Funct. Ecol. 8, 52–57.

Lebre, P.H., De Maayer, P., Cowan, D.A., 2017. Xerotolerant bacteria: surviving through a dry spell. Nat. Rev. Microbiol. 15 (5), 285–296.

Lee, K.C., et al., 2016. Niche filtering of bacteria in soil and rock habitats of the Colorado Plateau Desert, Utah, USA. Front. Microbiol. 7, 1–7.

Li, S., et al., 2013. Phylogenetic diversity of endolithic bacteria in Bole granite rock in Xinjiang. Acta Ecol. Sin. 33 (4), 178–184.

Luger, R., et al., 2017. A seven-planet resonant chain in TRAPPIST-1. Nat. Astron. 1, 129.

Makhalanyane, T.P., et al., 2013. Evidence of species recruitment and development of hot desert hypolithic communities. Environ. Microbiol. Rep. 5 (2), 219–224.

Makhalanyane, T.P., Valverde, A., Gunnigle, E., et al., 2015a. Microbial ecology of hot desert edaphic systems. FEMS Microbiol. Rev. 39 (2), 203–221.

Makhalanyane, T.P., Valverde, A., Velázquez, D., et al., 2015b. Ecology and biogeochemistry of cyanobacteria in soils, permafrost, aquatic and cryptic polar habitats. Biodivers. Conserv. 24 (4), 819–840.

Mares, M.A., 1999. In: Mares, M.A. (Ed.), Encyclopedia of Deserts. University of Oklahoma Press.

Matthes, U., Turner, S.J., Larson, D.W., 2001. Light attenuation by limestone rock and its constraint on the depth distribution of endolithic algae and cyanobacteria. Int. J. Plant Sci. 162 (2), 263–270.

McEwen, A., et al., 2011. Seasonal flows on warm Martian slopes. Science 333 (6043), 740–743.

McNamara, C.J., et al., 2006. Epilithic and endolithic bacterial communities in limestone from a Maya archaeological site. Microb. Ecol. 51 (1), 51–64.

Meslier, V., et al., 2018. Fundamental drivers for endolithic microbial community assemblies in the hyperarid Atacama Desert. Environ. Microbiol. 20, 1765–1781.

Michalski, G., Bohlke, J.K., Thiemens, M., 2004. Long term atmospheric deposition as thesource of nitrate and other salts in the Atacama Desert, Chile: new evidence frommass-independent oxygen isotopic compositions. Geochim. Cosmochim. Acta 68, 4023–4038.

Mohammadipanah, F., Wink, J., 2016. Actinobacteria from arid and desert habitats: diversity and biological activity. Front. Microbiol. 6, 1–10.

Mustard, J.F., et al., 2008. Hydrated silicate minerals on Mars observed by the Mars reconnaissance orbiter CRISM instrument. Nature 454 (7202), 305–309.

Navarro-Gonzalez, R., 2003. Mars-like soils in the Atacama Desert, Chile, and the dry limit of microbial life. Science 302 (5647), 1018–1021.

Nienow, J., 2009. Extremophiles: dry environments (including Cryptoendoliths). In: Encyclopedia of Microbiology. Elsevier, Oxford, pp. 159–173.

Nienow, J.A., McKay, C., Friedmann, E.I., 1988. The cryptoendolithic microbial environment in the Ross Desert of Antarctica: light in the photosynthetically active region. Microb. Ecol. 16 (3), 271–289.

Niles, P.B., et al., 2010. Stable isotope measurements of Martian atmospheric CO_2 at the Phoenix landing site. Science 329, 1334–1338.

Omelon, C.R., 2008. Endolithic microbial communities in Polar Desert habitats. Geomicrobiol J. 25 (7–8), 404–414.

Omelon, C.R., 2016. Endolithic microorganisms and their habitats. In: Hurst, C.J. (Ed.), Their World: A Diversity of Microbial Environments. Springer, pp. 171–201.

Osterloo, M., et al., 2008. Chloride-bearing materials in the southern highlands of Mars. Science 319, 1651–1654.

Palmer, R.J., Friedmann, E.I., 1990. Water relations and photosynthesis in the cryptoendolithic microbial habitat of hot and cold deserts. Microb. Ecol. 19 (1), 111–118.

Pointing, S.B., Belnap, J., 2012. Microbial colonization and controls in dryland systems. Nat. Rev. Microbiol. 10 (9), 654.

Pointing, S.B., et al., 2009. Highly specialized microbial diversity in hyper-arid polar desert. Proc. Natl. Acad. Sci. U. S. A. 106 (47), 19964–19969.

Pointing, S.B., et al., 2015. Biogeography of photoautotrophs in the high polar biome. Front. Plant Sci. 6, 692.

Pontefract, A., et al., 2014. Impact-generated endolithic habitat within crystalline rocks of the Haughton impact structure, Devon Island, Canada. Astrobiology 14 (6), 522–533.

Potts, M., 1999. Mechanisms of desiccation tolerance in Cyanobacteria. Eur. J. Phycol. 34 (4), 319–328.

Raanan, H., et al., 2015. Three-dimensional structure and cyanobacterial activity within a desert biological soil crust. Environ. Microbiol. 18 (2), 372–383.

Raven, J.A., Kübler, J.E., Beardall, J., 2000. Put out the light, and then put out the light. J. Mar. Biol. Assoc. U. K. 80 (2000), 1–25.

Reiss, D., et al., 2004. Absolute dune ages and implications for the time of formation of gullies in Nirgal Vallis, Mars. J. Geophys. Res. 109 (6), 1–9.

Robinson, C.K., et al., 2015. Microbial diversity and the presence of algae in halite endolithic communities are correlated to atmospheric moisture in the hyper-arid zone of the Atacama Desert. Environ. Microbiol. 17 (2), 299–315.

Seager, S., 2013. Exoplanet habitability. Science 340 (6132), 577–581.

Siebert, J., Hirsch, P., 1988. Characterization of 15 selected coccal bacteria isolated from Antarctic rock and soil samples from the McMurdo-dry valleys (South-Victoria land). Polar Biol. 9, 37–44.

Smith, H.D., et al., 2014. Comparative analysis of cyanobacteria inhabiting rocks with different light transmittance in the Mojave Desert: a Mars terrestrial analogue. Int. J. Astrobiol. 13 (3), 1–7.

Squyres, S.W., et al., 2004. The opportunity Rover's Athena science investigation at Meridiani Planum, Mars. Science 306 (5702), 1698–1703.

Tang, Y., et al., 2012. Endolithic bacterial communities in dolomite and limestone rocks from the Nanjiang canyon in Guizhou karst area (China). Geomicrobiol J. 29 (3), 213–225.

Tracy, C.R., et al., 2010. Microclimate and limits to photosynthesis in a diverse community of hypolithic cyanobacteria in northern Australia. Environ. Microbiol. 12 (3), 592–607.

Uritskiy, G., Getsin, S., Munn, A., Gomez-Silva, B., Davila, A., Glass, B., Taylor, J., DiRuggiero, J., 2019. Resilience and adaptation mechanisms of an extremophile microbial community following a catastrophic climate event. In: bioRxiv 442525.https:/doi.org/10.1101/442525.

Villar, S.E.J., Edwards, H.G.M., Cockell, C.S., 2005. Raman spectroscopy of endoliths from Antarctic cold desert environments. Analyst 130 (2), 156–162.

Vítek, P., et al., 2012. The miniaturized Raman system and detection of traces of life in halite from the Atacama Desert: some considerations for the search for life signatures on Mars. Astrobiology 12 (12), 1095–1099.

Vítek, P., et al., 2013. Phototrophic community in gypsum crust from the Atacama Desert studied by Raman spectroscopy and microscopic imaging. Geomicrobiol J. 30 (5), 399–410.

Vítek, P., et al., 2016. Raman imaging in geomicrobiology: endolithic phototrophic microorganisms in gypsum from the extreme sun irradiation area in the Atacama Desert. Anal. Bioanal. Chem. 408, 4083–4092.

Walker, J.J., Pace, N., 2007. Endolithic microbial ecosystems. Ann. Rev. Microbiol. 61, 331–347.

Wang, J., et al., 2013. Phylogenetic beta diversity in bacterial assemblages across ecosystems: deterministic versus stochastic processes. ISME J. 7, 1310–1321.

Warren-Rhodes, K.a., et al., 2006. Hypolithic cyanobacteria, dry limit of photosynthesis, and microbial ecology in the hyperarid Atacama Desert. Microb. Ecol. 52 (3), 389–398.

Warren-Rhodes, K., et al., 2007. Cyanobacterial ecology across environmental gradients and spatial scales in China's hot and cold deserts. FEMS Microbiol. Ecol. 61 (3), 470–482.

Warren-Rhodes, K.A., et al., 2013. Physical ecology of hypolithic communities in the central Namib Desert: the role of fog, rain, rock habitat, and light. J. Geophys. Res.: Biogeosci. 118 (4), 1451–1460.

Wei, S.T.S., et al., 2016. Taxonomic and functional diversity of soil and hypolithic microbial communities in Miers Valley, McMurdo Dry Valleys, Antarctica. Frontiers 7, 1–11.

Westall, F., Cavalazzi, B., 2011. Biosignatures in rocks. In: Thiel, V., Reitner, J. (Eds.), Encyclopedia of Geobiology. Springer, Berlin Heidelberg, pp. 189–201.

Wierzchos, J., Ascaso, C., 1994. Application of back-scattered electron imaging to the study of the lichen-rock interface. J. Microsc. 175 (1), 54–59.

Wierzchos, J., Ascaso, C., McKay, C.P., 2006. Endolithic cyanobacteria in halite rocks from the hyperarid core of the Atacama Desert. Astrobiology 6 (3), 415–422.

Wierzchos, J., et al., 2011. Microbial colonization of Ca-sulfate crusts in the hyperarid core of the Atacama Desert: implications for the search for life on Mars. Geobiology 9 (1), 44–60.

Wierzchos, J., De Los Ríos, A., Ascaso, C., 2012a. Microorganisms in desert rocks: the edge of life on Earth. Int. Microbiol. 15 (4), 173–183.

Wierzchos, J., et al., 2012b. Novel water source for endolithic life in the hyperarid core of the Atacama Desert. Biogeosciences 9 (6), 2275–2286.

Wierzchos, J., et al., 2013. Ignimbrite as a substrate for endolithic life in the hyper-arid Atacama Desert: implications for the search for life on Mars. Icarus 224 (2), 334–346.

Wierzchos, J., et al., 2015. Adaptation strategies of endolithic chlorophototrophs to survive the hyperarid and extreme solar radiation environment of the Atacama Desert. Front. Microbiol. 6, 1–11.

Zablocki, O., Adriaenssens, E.M., Cowan, D., 2015. Diversity and ecology of viruses in hyperarid desert soils. Appl. Environ. Microbiol. 82 (3), 770–777.

Ziolkowski, L.A., et al., 2013. Arctic gypsum endoliths: a biogeochemical characterization of a viable and active microbial community. Biogeosciences 10 (11), 7661–7675.

Zsom, A., et al., 2013. Toward the minimum inner edge distance of the habitable zone. Astrophys. J. 778 (2), 109.

Survival of subsurface microbial communities over geological times and the implications for astrobiology

Helga Stan-Lotter

Department of Microbiology, Division of Molecular Biology, University of Salzburg, Salzburg, Austria

Chapter outline

1 Introduction

Life can thrive in very harsh environments on Earth which were thought for a long time to be uninhabitable, but certain organisms—termed extremophiles—prefer unusual physicochemical factors such as very high or low temperatures, high or low pH values, high concentrations of salt or other chemicals, high pressure and others in their surroundings (Rothschild and Mancinelli, 2001; Seckbach, 2000;

Model Ecosystems in Extreme Environments. https://doi.org/10.1016/B978-0-12-812742-1.00008-8

Rampelotto, 2013; Seckbach et al., 2013; Kallmeyer and Wagner, 2014). Such capabilities are of interest to the search for extraterrestrial life, particularly in light of the recent explosion in the number of potentially habitable exoplanets and exomoons detected by advanced technologies (https://exoplanets.nasa.gov/news/, accessed February 2017). But conditions on the surfaces of those celestial bodies may be too extreme with respect to ionizing radiation, temperature, and availability of water and/or nutrients for supporting a biosphere. Information from the study of Mars suggested that—considering a similar geological past like Earth—if life has been developed also on Mars it may more likely be found in the subsurface of the planet. Plans for drilling into various depths of the Martian subsurface were developed in a NASA workshop in 1998 (published by Mancinelli, 2000), and sophisticated procedures as well as refined technical equipment were presented recently (McKay et al., 2013; Zacny et al., 2013). A chapter by Cockell (2014) considers in detail the existing knowledge about the habitability of the Martian subsurface and makes comparisons with the terrestrial subsurface and analog environments.

The exploration of microbial life in subterranean environments, such as deep subsea floor sediments, crustal rocks, sedimentary rocks, and ancient salt deposits has increased substantially in recent years. One reason was the provocative article by Whitman et al. (1998), who estimated that the total amount of carbon in the "intraterrestrial" prokaryotic mass on Earth may be as large or even exceed that of plants and prokaryotes growing on the surface of the Earth. Even if newer assessments have challenged the magnitude of the proposed biomass (Kallmeyer et al., 2012; Jørgensen, 2012), many scientists believe that indeed subsurface life may be the main stage for the planet's biodiversity (Whitman et al., 1998). There is also interest in the subsurface biosphere from application-oriented activities, such as oil exploration, waste storage issues, use of hydrothermal systems, and others (see Kallmeyer and Wagner, 2014).

In this chapter the extremely halophilic portion of the subterranean biomass is considered, as far as it is accessible in some of the ancient salt sediments of the world. Current notions on marine sediments and the longevity of their microbial communities are presented.

2 Ancient salt deposits

During several periods in Earth history, huge sedimentation of halite and other minerals from hypersaline seas took place. Such deposits are found predominantly in the Northern regions of the continents, for example, in Siberia, Canada (Mackenzie basin), Northern and Central Europe (Zechstein series), South-Eastern Europe (Alps and Carpathian mountains), and the Midcontinent basin in North America (Zharkov, 1981). The distribution of perceived Permian salt deposits is shown in Fig. 1. The continental land masses at the time of deposition were concentrated around the paleoequator and formed the supercontinent Pangaea. The climate was arid and dry, giving rise to large-scale evaporation. About 200 million years ago, Pangaea

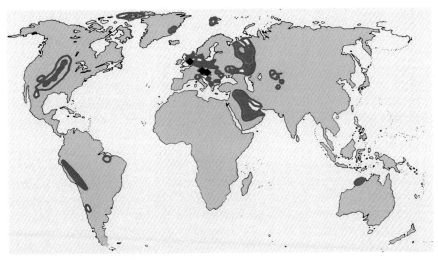

FIG. 1

Distribution of known and presumed Permian salt sediments (modified from Javor, 1989). Black diamonds indicate locations of rock salt samples (Zechstein deposit in England; Alpine deposits in Germany and Austria), from which viable haloarchaea were isolated. From Fendrihan et al. (2006), with permission.

was starting to break up; the land masses were moving laterally as well as to the North; and mountain ranges such as the Alps, the Carpathians, and the Himalayas were pushed up by plate tectonics (http://science.jrank.org/pages/1751/Continental-Drift-Pangaea-splits.html, accessed February 20, 2017; Einsele, 1992; Erickson, 2001). Some salt deposits are therefore found at high altitudes, for example, in the Alps at 500–1200 m. They were overlain by carbonate, limestone, clay, and other materials, which formed up to the Early Cretaceous age (c. 100–145 million years ago; Spötl and Hasenhüttl, 1998). Tectonic forces often caused deformations of layers and interspersing with large rock accumulations (see Radax et al., 2001). The clay and carbonate layers largely prevented the washing out of halite during the heavy precipitations of the ice ages. Alpine salt deposits vary in thickness from 250 to 700 m, although some deeply buried layers were estimated to be up to 1000 m. They contain up to 97% halite, some anhydrite, polyhalite, and traces of iron and other metal ions (Schauberger, 1986). In the Alpine basin and in the region of the Zechstein Sea, which covered Northern Europe, no more salt sedimentation took place after the Triassic period.

In contrast to other sediments, salt deposits are generally devoid of macroscopic fossils on which an age determination could be based. Instead, palynological and isotope studies, in addition to stratigraphical information, were applied. Klaus (1955, 1974) detected in dissolved rock salt from Alpine halite deposits plant spores from extinct species, which exhibited well preserved morphological features. The spores

Pityosporites, Gigantosporites, and others are characteristic for the Permian period and can be clearly distinguished from *Triadosporites* species, which are found in Triassic evaporites. The formation of most of these Alpine salt sediments and the Zechstein deposits were thus dated to the Upper Permian period, while some Alpine deposits were dated to the Triassic period.

Sulfur isotope ratios (expressed as $\delta^{34}S$) are used for evaporites which contain sulfates of marine origin (Holser and Kaplan, 1966). Each time period can be characterized by its $\delta^{34}S$ value, generally represented by evaporites from a variety of geographic locations (Claypool et al., 1980). Worldwide results from samples of doubtless stratigraphical relationships showed an extremely low $\delta^{34}S$ value for evaporites of Permian age (+8 to +12‰), and a higher $\delta^{34}S$ value (+20 to +27‰) for those of Triassic origin. The factors that influence $\delta^{34}S$ are thought to include reduction of seawater sulfate by bacteria and its fixation as pyrite in muds, oxidative weathering of old pyrite and its return to the sea as sulfate, crystallization of sulfate minerals (mainly gypsum or anhydrite) in evaporites, and the weathering of old evaporites and return to the sea, as outlined by Claypool et al. (1980). Using sulfur isotope ratios from numerous anhydrite and gypsum samples, Pak and Schauberger (1981) could confirm a Permian or a Triassic age for many of the Alpine salt deposits ($\delta^{34}S$ values of +10 or +25/27‰, respectively; Table 1). Likewise, the Zechstein series of deposits were corroborated to be of Permian origin as well as the Midcontinent basin in the United States (Table 1), on the basis of closely corresponding $\delta^{34}S$ values. These data were considered a particularly strong affirmation of the palynological results by Klaus (1955, 1974), indicating an Upper Permian age for the Alpine red salt deposits (Schauberger, 1986).

As mentioned, macrofossils are not known to be contained in halite. However, microfossils have been found. Besides plant spores and pollen grains, cellulose fibers were identified in the 250-million-year-old Salado formation in southeastern New Mexico, USA (Griffith et al., 2008).

Table 1 Dating of saline deposits and rocks of marine origin by determination of sulfur isotopes (Pak and Schauberger, 1981; Holser and Kaplan, 1966)

Salt deposit	$\delta^{34}S$ (‰)	Age (million years)
Altaussee (Austria)	+10.5 to +11.7	Upper Perm (270–251)
Bad Ischl (Austria), red	+10.4 to +11.8	Upper Perm (270–251)
Hallstatt (Austria)	+10.1 to +11.8	Upper Perm (270–251)
Berchtesgaden (Germany)	+10.9 to +11.3	Upper Perm (270–251)
Zechstein layer (Germany)	+8 to +12	Perm (257–240)
Midcontinent basin (USA)	+11.5 ± 1[a]	Perm (257–240)
Bad Ischl (Austria), gray	+26.7	Trias/Skyth (220–200)
Gypsum Niedernhall (Germany)	+20.6 to +21.6[b]	Trias (230–200)

[a]*Thode and Monster (1965).*
[b]*Nielsen and Ricke (1964).*

Even more interesting are "viable fossils," such as haloarchaea and halophilic bacteria which were found in ancient salt deposits and are treated in the following sections.

---

3 Haloarchaea

Evaporation of seawater or saline lakes creates surface waters with NaCl concentrations approaching saturation. These contain communities of extremely halophilic microorganisms consisting of mainly of archaea (now termed haloarchaea), bacteria, and unicellular eukaryotes. They are found in hypersaline surface waters such as the Dead Sea, the Great Salt Lake, and natural or artificial salterns (Javor, 1989). Haloarchaea make their presence known by their intense red, pink, or purple pigmentation, which is due to carotenoids in their membranes, to the C_{50} compound bacterioruberin or its derivatives (Kamekura, 1993), and in some species, due to the retinal-containing protein bacteriorhodopsin (Javor, 1989). Extremely halophilic microorganisms balance the high osmotic pressure of the surrounding brines by concentrating KCl in their cytoplasm (up to 100 times higher than in the environment) or by accumulation of compatible solutes (Oren, 2008). In the laboratory, haloarchaea grow optimally in media containing 2.5–5.2 M NaCl (Kushner and Kamekura, 1988). Most haloarchaea except the coccoid genera lyse quickly when exposed to salt concentrations of less than 10%, but some exceptions are known (Oren, 2008). At present, haloarchaeal strains are classified into 53 genera with 215 species (as of February 2017; see http://www.bacterio.net).

When saline waters evaporate, sodium chloride crystals form and haloarchaea can get included into those crystals. In laboratory experiments they are still visible by their color and their motility can be observed by microscopy for some weeks after inclusion (Norton and Grant, 1988). The ancient evaporites originated from such crystals, and it is conceivable that haloarchaea survived in the fluid inclusions which form in halite during crystal growth (Roedder, 1984). Prestaining of haloarchaea with the fluorescent dye SYTO9 and subsequent embedding in salt by drying the cellular suspensions suggested that cells were only present in fluid inclusions and not in the crystal matrix of halite (Fig. 2). A similar observation had been made with fluorescent *Pseudomonas* cells which were enclosed in laboratory-grown halite (Adamski et al., 2006). The bright green fluorescence of stained cells outlined the morphology of the characteristic rectangular fluid inclusions of halite (Fig. 2, left panel). At higher magnifications, individual cells became visible (Fig. 2, right panel) and most of them appeared of roundish morphology.

There have been several reports of viable halophilic prokaryotes being recovered from geologically old halite deposits and brought to proliferation (e.g., Grant et al., 1998; McGenity et al., 2000; Stan-Lotter et al., 1999; Stan-Lotter et al., 2002; Gruber et al., 2004; Vreeland et al., 2007; Schubert et al., 2010; Gramain et al., 2011; Jaakkola et al., 2014). The majority of the isolates were haloarchaea, but some bacteria were also recovered (Jaakkola et al., 2016). In addition, haloarchaeal 16S rDNA

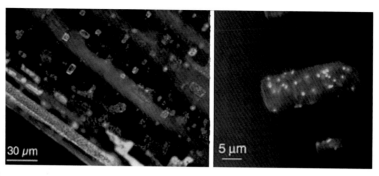

FIG. 2

Localization of prestained haloarchaea in fluid inclusions. Cells were stained with SYTO9 prior to embedding in artificial halite. Low (left panel) and high (right panel) magnification of *Halobacterium salinarum* NRC-1 cells, embedded in fluid inclusions for about 2 days. Cells were observed with a Zeiss Axioskop fluorescence microscope.

Photographs taken by S. Fendrihan.

sequences were obtained from rock salt samples (Radax et al., 2001; Fish et al., 2002; Park et al., 2009; Gramain et al., 2011; Sankaranarayanan et al., 2014).

The repeated and independent isolation of viable haloarchaea from ancient evaporites in several parts of the world, together with the recovery of amplifiable DNA—sometimes from the same samples—suggested that the microorganisms might be of the same age as the sediments and were originally entrapped when the ancient brines dried down. How they survived geological time periods is not known and currently a focus of research. Several mechanisms which were put forward recently and may contribute to a plausible explanation of prokaryotic longevity are reviewed in the next section.

4 Mechanisms for microbial and haloarchaeal longevity

Endospores which are produced by certain low G + C Gram-positive bacteria can survive environmental assaults that would normally kill bacteria, such as high temperature, high UV irradiation, desiccation, chemical damage, and enzymatic destruction. A variety of different microorganisms form spores or cysts, but the endospores of Gram-positive bacteria are by far the most resistant to environmental extremes on Earth and can show extreme longevity (Nicholson et al., 2000). Reliable reports exist of the recovery and revival of spores from environmental samples as old as 10^5 years, for example, lake sediments (Nicholson et al., 2000) and, although more controversial, from the gut of a bee fossilized in amber for an estimated 25–40 million years (Cano and Borucki, 1995).

However, genome sequences suggested that haloarchaea—or archaea in general—do not produce spores (Onyenwoke et al., 2004). The occurrence of other types of haloarchaeal resting states or dormant forms, such as cysts, has been discussed (Grant et al., 1998), but was not yet demonstrated unequivocally. So which mechanisms might be involved in their (potential) longevity?

The following findings were suggested to promote longevity:

- Oligotrophy, starvation, and dwarfing.
- Desiccation and radiation resistance.
- DNA as phosphate source.
- Polyploidy.
- Formation of spheres.

Some mechanisms are known from other members in the prokaryotic world, some are novel and appear—at present—unique to extremely halophilic Archaea.

4.1 Oligotrophy, starvation, and dwarfing

Oligotrophs are organisms that can live in nutrient-poor environments. Typical are slow growth and low rates of metabolism, which lead generally to low population density. Oligotrophs occur in deep oceanic sediments, caves, glacial and polar ice, deep subsurface environments, ocean waters, and leached soils. The concentration of total organic carbon in these environments is typically in the range of one to a few milligrams per liter (Egli, 2010). Adaptation to nutrient limitation can consist in increasing the surface-to-volume ratio which increases the capacity for nutrient uptake relative to cell volume. Most oligotrophs achieve a high surface-to-volume ratio by reducing their cell diameter and forming miniaturized cells. In this way the organism's capacity to scavenge available energy-yielding substrates will be increased (Morita, 1982). A high surface-to-volume ratio is thus typical for many oligotrophic bacteria.

Reducing the cell diameter upon exposure to starvation was observed and has been called "dwarfing" (Kjelleberg et al., 1983; Nyström, 2004). The dwarf status is reversed to growth as rod-shaped or pleomorphic cells, once the conditions for proliferation are met. So far, several bacterial dwarfs have been described, but also an archaeal dwarf was identified in arid soil (Rutz and Kieft, 2004). There is a limit to miniaturization, which is due to the required minimum contents of DNA, ribosomes, RNA, and proteins. The theoretical prediction of the minimum volume is $0.005–0.01\ \mu m^3$, which was nearly reached by starved cells of the marine bacterium *Candidatus* Pelagibacter ubique, a member of the SAR11 clade (Steindler et al., 2011). A special mode of dwarfing appears to occur in *Halobacterium* and other haloarchaea as response to reduced water activity (a_w) in their surroundings (see below Section 4.4).

Dormant or starved bacteria and spores contain reduced amounts of ATP, for example, the ATP content of dormant spores of several *Bacillus* species was about two orders of magnitude lower than that of actively growing *Bacillus* cells (Setlow and Kornberg, 1970). Marine *Vibrio* species, which are deemed dormant when in the oceans, lost ATP during starvation experiments (Oliver and Stringer, 1984), and several strains of *Staphylococcus* showed a decline of ATP when adhering to polymer surfaces and entering a dormant, miniaturized state (Stollenwerk et al., 1998).

4.2 Desiccation and radiation resistance

The resistance to ionizing radiation of the haloarchaeon *Halobacterium salinarum* is quite remarkable. It showed a D_{10} value (the dose of radiation in Gray (Gy) that reduces the survival of a population by 90%) of 5 kGy, which is of a similar order of magnitude as that of the archaeon *Thermococcus gammatolerans* (6 kGy) and of the bacterium *Deinococcus radiodurans* (12 kGy) (Kottemann et al., 2005; Webb and DiRuggiero, 2013). Two irradiation-derived mutants of *Hbt. salinarum* displayed even D_{10} values close to that of *D. radiodurans* (DeVeaux et al., 2007). Both desiccation and irradiation result in double-strand breaks of DNA. The connection between resistance to these treatments had been established previously (Daly, 2009). In *H. salinarum* recovery from irradiation as well as desiccation occurred quickly—within hours of damage induction (Kottemann et al., 2005). An explanation for the mechanism underlying this behavior is the presence of several chromosomes and the ability to regenerate intact chromosomes from scattered fragments (Soppa, 2013). Overlapping genomic fragments are a prerequisite for this mechanism, and thus, it only operates in oligoploid species such as *D. radiodurans*, or polyploid species, such as haloarchaea (Soppa, 2014); see later.

Natural halite contains considerable background radiation from the isotope ^{40}K, which has been suggested to limit the viability of spores in fluid inclusions to a maximum of 109 million years or less (Kminek et al., 2003). However, in light of the polyploidy of many haloarchaea and other prokaryotes this notion may be valid for monoploid species, but is not applicable to polyploid species (Soppa, 2014). As long as maintenance metabolism is possible, DNA damage in some copies of the chromosome can be repaired using the remaining intact copies of polyploid species as templates (Soppa, 2013, 2014).

4.3 DNA as phosphate source

Several groups of marine bacteria can grow on external dissolved DNA (Lennon, 2007) using it as a source of carbon, nitrogen, and phosphorus. Recently, it was shown that the haloarchaeon *Haloferax volcanii* is also able to use external DNA as a nutrient source and even internal DNA as a source of phosphate (Zerulla et al., 2014; Zerulla and Soppa, 2014). During phosphate starvation cells multiplied by degrading their chromosomes. The latter ability is possible due to its polyploid state (see later). That DNA has roles other than storing genetic information was, for example, deduced by its stabilizing function in biofilms (Zerulla et al., 2014). The findings have implications on the survival of haloarchaea in fluid inclusions. Microscopic algae, such as *Dunaliella* sp., were detected in fluid inclusions, together with haloarchaea (Schubert et al., 2010). Their lysis products, probably together with those of haloarchaea containing DNA and other molecules, might be sufficient to sustain a minimal maintenance metabolism, including DNA repair, of intact cells.

4.4 Polyploidy

Bacteria were generally assumed to be monoploid, with perhaps some exceptions such as *D. radiodurans*, which possesses 5–8 genome copies and is thus termed oligoploid (Zerulla and Soppa, 2014). The studies by Soppa and coworkers indicated that many prokaryotes are oligoploid or polyploid (10 or more genome copies), and most have a higher copy number during exponential growth phase than in stationary phase (Zerulla and Soppa, 2014; Griese et al., 2011; Soppa, 2011; Breuert et al., 2006). Several species of haloarchaea have been shown to be polyploid in all growth phases and at various growth rates, for example, *Halobacterium cutirubrum, H. salinarum, H. volcanii, Haloferax mediterranei*, and several new isolates (Zerulla and Soppa, 2014). Until now, no haloarchaeal species has been found to be monoploid; therefore, it might be that polyploidy is a general trait of haloarchaea, although more haloarchaeal species should be tested before a generalization is possible (Zerulla and Soppa, 2014). The molecular mechanisms of genome copy number regulation have not yet been unraveled.

Several positive aspects and evolutionary advantages of haloarchaeal polyploidy were discussed by Soppa (2013) and Zerulla and Soppa (2014). In the context of this article the considerations of survivability in the subsurface and especially in fluid inclusions are of interest. Soppa (2014) pointed out that the discrepancy between the predicted age of the isolates from ancient salt deposits and the instability of DNA over long times would be solved if haloarchaea surviving over geological times would be polyploid. They could accumulate many double-strand breaks and would be able to reconstruct intact chromosomes from scattered pieces, as has been described for *D. radiodurans* (Soppa, 2014). Maintenance energy could originate from dead and lysed cells that were originally also entrapped in the same fluid inclusion, as mentioned before (Section 3.3).

Interestingly, three haloarchaeal species have been isolated recently from 38 to 41 million-year-old salt deposits, and all three were shown to be polyploid, in agreement with the prediction that polyploidy enables long-term survival (Jaakkola et al., 2014).

4.5 Haloarchaeal sphere formation

The presence of small spherical particles with less than 1 μm diameter in fluid inclusions of 22,000–34,000-year-old halite was reported by Schubert et al. (2010). From such spheres cultures of the haloarchaeal genera *Halorubrum, Natronomonas*, and *Haloterrigena* were obtained, proving the viability of haloarchaeal cells in the ancient halite.

A transformation of rod-shaped haloarchaea to spherical particles was observed in laboratory-grown salt crystals (see Fig. 2B; Fendrihan et al., 2004). We could show that sphere formation is apparently a response to low external water activity (a_w) of several haloarchaeal species (Fendrihan et al., 2012). From one rod-shaped haloarchaeon, three or four spherical particles emerged (Fig. 3), which regrew to normal rods in nutrient media. The biochemical properties of rods and spheres were similar, except for an about 50-fold reduced content of ATP and a prolonged lag phase of spheres, thus bearing some resemblance to bacterial spores (Fendrihan et al., 2012). An indirect confirmation of polyploidy in haloarchaea (see Section 4.4) was provided by the culturability of several spheres which originated from one rod, since all spheres had received at least one complete genome.

The formation of spheres described here yields roundish mini-cells from several haloarchaeal species (Fendrihan et al., 2012), which resemble in size and shape the small particles from a Death Valley salt core termed "miniaturized cocci" by Schubert et al. (2010). Many questions still need to be resolved, for example, the mechanism of apparent multiple fission, which is used by the haloarchaeal cells,

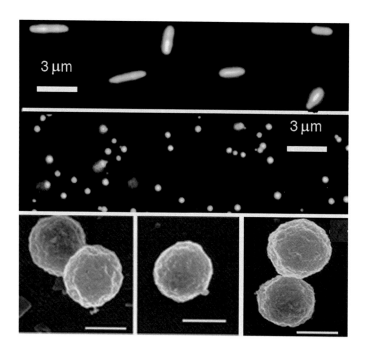

FIG. 3

Formation of haloarchaeal spheres. Rod-shaped cells of *Halobacterium salinarum* NRC-1 stained with SYTO9 (upper panel) were converted to spherical particles (middle panel) following exposure to a_w of less than 0.75 (Fendrihan et al., 2012). Scanning electron micrographs of spheres (lower panels). Spheres had formed within fluid inclusions of laboratory-grown halite and were obtained after dissolution of salt crystals. Bars in lower panel, 270 nm.

Fluorescent micrographs taken by S. Fendrihan, SEM photographs taken by C. Frethem.

or the influence of reduced a_w on surface proteins and other molecules as well as the role of cell components, which are released during the transit from rods to spheres (Fendrihan et al., 2012).

5 Extraterrestrial halite

Extraterrestrial halite has been identified in Martian meteorites (Treiman et al., 2000), in the Murchison and other carbonaceous meteorites (Barber, 1981) and in the Monahans meteorite, together with sylvite (KCl) and water inclusions (Zolensky et al., 1999). Postberg et al. (2009) found that about 6% of the ice grains from the plumes of Saturn's moon Enceladus are containing roughly 1.5% of a mixture of sodium chloride, sodium carbonate, and sodium bicarbonate. Recent images from the Mars Reconnaissance Orbiter showed evidence for seasonal emergence of liquid flows down steep rocky cliffs in summer, termed Recurring Slope lineae (RSL), which would be consistent with briny liquid water emerging from underground reservoirs on Mars (McEwen et al., 2011; Ojha et al., 2015). There is evidence for various brines on Jupiter's moon Europa that are composed primarily of water and salts (Muñoz-Iglesias et al., 2013). All of these discoveries make the consideration of potential habitats for halophilic life in space intriguing.

As one example, Enceladus is considered here because it is of special interest and has even been called "the best (current) astrobiology target in the Solar System" (McKay et al., 2014). Enceladus is Saturn's sixth largest moon, only 252 km in mean radius. Hydrothermal vents spew water vapor and ice particles from an underground ocean beneath the icy crust of Enceladus (https://saturn.jpl.nasa.gov/science/enceladus/, accessed February 2017; Fig. 4). These plumes include organic compounds, volatile gases, carbon dioxide, carbon monoxide, and as mentioned various salts. With its global ocean, unique chemistry and internal heat, Enceladus has become a promising lead in the search for worlds where life could exist. Tsou et al. (2012) suggested a mission called LIFE (Life Investigation For Enceladus). They pointed out that acquisition of samples would be comparatively easy, since in a low cost fly-by mission sufficient material from the plumes could be obtained for analyses on Earth—no landing would be necessary, and the total mission time would be 13.5 years.

6 Other ancient environments with microbial communities: Marine sediments

6.1 Subsurface life under energy limitation

Microbial life is widespread in the marine sediments that cover more than two-thirds of Earth's surface (D'Hondt et al., 2004). Viable bacterial populations in the sediments of five Pacific Ocean sites to depths of more than 500 m were described by

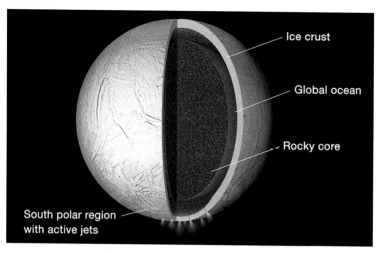

FIG. 4

Illustration of the interior of Saturn's moon Enceladus showing a global liquid water ocean (*dark blue*) between its rocky core and ice crust. The South polar region with active jets is indicated. Thickness of layers shown is not to scale.

Image credit: NASA/JPL-Caltech.

Parkes et al. (1994). Lipp et al. (2008) identified for the first time archaea as a significant portion of the microbial biomass in several marine sediments from around the world. Lomstein et al. (2012) pointed out the presence of large amounts of endospores in sediments, which were estimated to be 10 million years of age.

The extent and properties of the subsurface microbial communities are still widely unknown, probably due to severe energy limitation, starvation, and no-growth states, which are challenging for investigations (see review by Lever et al., 2015). Jørgensen and Marshall (2016) even state "The ability of large microbial communities to subsist under the extremely low energy and nutrient fluxes available in million year old sediments is one of the most enigmatic aspects of microbial life on our planet."

Energy limitation is probably the prevailing physiological state among microorganisms on earth (Morita, 1997), yet the theoretical energy minima required by microbial cells and the energy sources available to microbes in the subsurface environment are not well understood (Lever et al., 2015). Besides dwarfing as a response to starvation, which was mentioned before for haloarchaea and is characteristic of many marine subsurface microbes, other cellular changes such as cell shape, motility, and cell adhesion were identified and are described in detail in the review by Lever et al. (2015).

6.2 Dating microbial communities by amino acid racemization

The age of the marine sediments and of the inhabiting microbial communities are of special interest, since they are relevant to turnover times of carbon and other elements, living biomass, and also necromass (Lomstein et al., 2012;

Jørgensen and Marshall, 2016). Amino acid racemization which was introduced by Bada (1982) for dating of archaeological and geologically ancient materials is being used for these purposes. The method is based on the fact that once biologically synthesized amino acids are no longer maintaining their optical activity, that is, on the death of an organism, abiotic racemization reactions gradually convert the amino acids into an optically inactive mixture. The racemization rate of aspartic acid is the highest of all and therefore the D/L ratio of Asp is commonly determined. Although many factors influence racemization—predominantly the in situ temperature, but also the localization of amino acids in proteins, the pH, water content, and so on—and measurements of similar samples can yield significant differences, the method is widely applied, with progressively sophisticated corrections (Lomstein et al., 2012; Steen et al., 2013; Onstott et al., 2014).

The deep subsurface sediment material used by Lomstein et al. (2012) was obtained from the eastern tropical Pacific during the Ocean Drilling Program (ODP). Coring sites ranged from the continental shelf off the coast of Peru to ocean depths of 5000 m. Sediments at depths of up to 420 m below the sea floor (mbsf) were recovered and this sediment was found to be up to 35 million years old, as determined by carbon dating in the upper few meters and by stratigraphic correlation to other ODP sites (Lomstein et al., 2012; supplementary material). Application of amino acid racemization allowed to quantify the distributions and turnover times of living microbial biomass, endospores, and microbial necromass. For example, the total organic carbon turnover time was found to be about 43 million years, and it increased with approximately an order of magnitude from the upper and youngest sediment layers to the base of the drill cores. This suggested that buried organic carbon is indeed sufficient to fuel microbial activities over time scales of millions of years (Lomstein et al., 2012)—an important conclusion.

Referring back to microbes embedded in fluid inclusions of halite, it would be interesting if an attempt was made to determine the age of biological material in halite by the D/L racemization procedure, although the total organic carbon (TOC) there is much lower as in marine sediments (0.014%–0.057% in rock salt [Stan-Lotter, unpublished results] versus an average of 0.13% in habitable subsurface sediments [Lipp et al., 2008]).

7 Conclusions

The trustworthiness of the determination of geological ages in connection with biological issues (microbial longevity, dormancy, energy limitation) is increasing, since a high number of samples is being analyzed in the Ocean Drilling Programs, yielding a wealth of data relevant to the assessment of ancient microbial life.

Haloarchaea can survive extreme desiccation and starvation, sometimes for geological time periods, that is, millions of years. The mechanisms that appear to be involved include dwarfing of cells, reduction of ATP, and probably the production of dormant stages such as small spherical particles. Polyploidy is an essential factor in several of these strategies.

The apparent longevity of haloarchaeal strains in dry salty environments is relevant for astrobiological studies in general and, in particular, for the search for life on Mars, Europa, and Enceladus.

Acknowledgments

Thanks go to former coworkers and students for their contributions. Support of my laboratory work by the Austrian Science Foundation (FWF), grants P16260 and P18256, and the Austrian Research Promotion Agency (FFG), grant ASAP 815141, is gratefully acknowledged. The members of Huffman Laboratories, Golden, CO, USA, are thanked for their enthusiastic TOC determination of ancient rock salt.

References

Adamski, J.C., Roberts, J.A., Goldstein, R.H., 2006. Entrapment of bacteria in fluid inclusions in laboratory-grown halite. Astrobiology 6, 552–562.

Bada, J.L., 1982. Racemization of amino acids in nature. Interdiscip. Sci. Rev. 7, 30–46.

Barber, D.J., 1981. Matrix phyllosilicates and associated minerals in C2M carbonaceous chondrites. Geochim. Cosmochim. Acta 45, 945–970.

Breuert, S., Allers, T., Spohn, G., Soppa, J., 2006. Regulated polyploidy in halophilic archaea. PLoS ONE. 1. e92.

Cano, R.J., Borucki, M.K., 1995. Revival and identification of bacterial spores in 25- to 40-million-year-old Dominican amber. Science 268, 1060–1064.

Claypool, G.E., Holser, W.T., Kaplan, I.R., Sakai, H., Zak, I., 1980. The age curves of sulfur and oxygen isotopes in marine sulfate and their mutual interpretation. Chem. Geol. 28, 199–259.

Cockell, C.S., 2014. The subsurface habitability of terrestrial rocky planets: mars. In: Kallmeyer, J., Wagner, D. (Eds.), Microbial Life of the Deep Biosphere. Walter de Gruyter GmbH, Berlin/Boston, pp. 225–259.

Daly, M.J., 2009. A new perspective on radiation resistance based on *Deinococcus radiodurans*. Nat Rev Microbiol 7, 237–245.

DeVeaux, L.C., Müller, J.A., Smith, J.R., Petrisko, J., Wells, D.P., DasSarma, S., 2007. Extremely radiation-resistant mutants of a halophilic archaeon with increased single-stranded DNA-binding protein (RPA) gene expression. Radiat. Res. 168, 507–514.

D'Hondt, S., Jørgensen, B.B., Miller, D.J., Batzke, A., Blake, R., Cragg, B.A., Cypionka, H., Dickens, G.R., Ferdelman, T., Hinrichs, K.U., Holm, N.G., Mitterer, R., Spivack, A., Wang, G., Bekins, B., Engelen, B., Ford, K., Gettemy, G., Rutherford, S.D., Sass, H., Skilbeck, C.G., Aiello, I.W., Guèrin, G., House, C.H., Inagaki, F., Meister, P., Naehr, T., Niitsuma, S., Parkes, R.J., Schippers, A., Smith, D.C., Teske, A., Wiegel, J., Padilla, C.N., Acosta, J.L., 2004. Distributions of microbial activities in deep subseafloor sediments. Science 306, 2216–2221.

Egli, T., 2010. How to live at very low substrate concentration. Water Res. 44, 4826–4837.

Einsele, G., 1992. Sedimentary Basins. Evolution, Facies and Sediment Budget. Springer Verlag Berlin, Heidelberg, New York.

Erickson, J., 2001. Plate Tectonics. Unraveling the mysteries of the Earth, Revised edition. Facts on File Inc., New York.

Fendrihan, S., Leuko, S., Stan-Lotter, H., 2004. Effects of embedding *Halobacterium* sp. NRC-1 in salt crystals and potential implications for long term preservation. In: Proceedings of the Third European Workshop on Exo-Astrobiology, ESA SP-545, pp. 203–204.

Fendrihan, S., Legat, A., Pfaffenhuemer, M., Gruber, C., Weidler, G., Gerbl, F., Stan-Lotter, H., 2006. Extremely halophilic archaea and the issue of long term microbial survival. Rev. Environ. Sci. Biotechnol. 5, 203–218.

Fendrihan, S., Dornmayr-Pfaffenhuemer, M., Gerbl, F.W., Holzinger, A., Grösbacher, M., Briza, P., Erler, A., Gruber, C., Plätzer, K., Stan-Lotter, H., 2012. Spherical particles of halophilic archaea correlate with exposure to low water activity—implications for microbial survival in fluid inclusions of ancient halite. Geobiology 10, 424–433.

Fish, S.A., Shepherd, T.J., McGenity, T.J., Grant, W.D., 2002. Recovery of 16S ribosomal RNA gene fragments from ancient halite. Nature 417, 432–436.

Gramain, A., Chong Diaz, G.C., Demergasso, C., Lowenstein, T.K., McGenity, T.J., 2011. Archaeal diversity along a subterranean salt core from the Salar Grande (Chile). Environ. Microbiol. 13, 2105–2121.

Grant, W.D., Gemmell, R.T., McGenity, T.J., 1998. Halobacteria: the evidence for longevity. Extremophiles 2, 279–287.

Griese, M., Lange, C., Soppa, J., 2011. Ploidy in cyanobacteria. FEMS Microbiol. Lett. 323, 124–131. https://doi.org/10.1111/j.1574-6968.2011.02368.x.

Griffith, J.D., Willcox, S., Powers, D.W., Nelson, R., Baxter, B.K., 2008. Discovery of abundant cellulose microfibers encased in 250 Ma Permian halite: a macromolecular target in the search for life on other planets. Astrobiology 8, 215–228.

Gruber, C., Legat, A., Pfaffenhuemer, M., Radax, C., Weidler, G., Busse, H.-J., Stan-Lotter, H., 2004. *Halobacterium noricense* sp. nov., an archaeal isolate from a bore core of an alpine Permo-Triassic salt deposit, classification of *Halobacterium* sp. NRC-1 as a strain of *Halobacterium salinarum* and emended description of *Halobacterium salinarum*. Extremophiles 8, 431–439.

Holser, W.T., Kaplan, I.R., 1966. Isotope geochemistry of sedimentary sulfates. Chem. Geol. 1, 93–135.

Jaakkola, S.T., Zerulla, K., Guo, Q., Liu, Y., Ma, H., Yang, C., Bamford, D.H., Chen, X., Soppa, J., Oksanen, H.M., 2014. Halophilic archaea cultivated from surface sterilized middle-late Eocene rock salt are polyploid. PLoS ONE. 9, e110533.

Jaakkola, S.T., Ravantti, J.J., Oksanen, H.M., Bamford, D.H., 2016. Buried alive: Microbes from ancient halite. Trends Microbiol. 24, 148–160. https://doi.org/10.1016/j.tim.2015.12.002.

Javor, B.J., 1989. Hypersaline Environments: Microbiology and Biogeochemistry. Springer Verlag, Berlin, Heidelberg, New York.

Jørgensen, B.B., 2012. Shrinking majority of the deep biosphere. Proc. Natl. Acad. Sci. U. S. A. 109, 15976–15977. https://doi.org/10.1073/pnas.1213639109.

Jørgensen, B.B., Marshall, I.P., 2016. Slow microbial life in the seabed. Annu. Rev. Mar. Sci. 8, 311–332. https://doi.org/10.1146/annurev-marine-010814-015535.

Kallmeyer, J., Wagner, D. (Eds.), 2014. Microbial Life of the Deep Biosphere. Walter de Gruyter GmbH, Berlin/Boston.

Kallmeyer, J., Pockalny, R., Adhikari, R.R., Smith, D.C., D'Hondt, S., 2012. Global distribution of microbial abundance and biomass in subseafloor sediment. Proc. Natl. Acad. Sci. U. S. A. 109, 16213–16216.

Kamekura, M., 1993. Lipids of extreme halophiles. In: Vreeland, R.H., Hochstein, L.I. (Eds.), The Biology of Halophilic Bacteria. CRC Press, Boca Raton, pp. 135–161.

Kjelleberg, S., Humphrey, B.B., Marshall, K.C., 1983. Initial phases of starvation and activity of bacteria at surfaces. Appl. Environ. Microbiol. 46, 978–984.

Klaus, W., 1955. Über die Sporendiagnose des deutschen Zechsteinsalzes und des alpinen Salzgebirges. Z. Dtsch. Geol. Ges. 105, 756–788.

Klaus, W., 1974. Neue Beiträge zur Datierung von Evaporiten des Oberperm. Carinthia II, 164. Jahrg 84, 79–85.

Kminek, G., Bada, J.L., Pogliano, K., Ward, J.F., 2003. Radiation-dependent limit for the viability of bacterial spores in halite fluid inclusions and on Mars. Radiat. Res. 159, 722–729.

Kottcmann, M., Kish, A., Iloanusi, C., Bjork, S., DiRuggiero, J., 2005. Physiological responses of the halophilic archaeon *Halobacterium* sp. strain NRC-1 to desiccation and gamma irradiation. Extremophiles 9, 219–227.

Kushner, D.J., Kamekura, M., 1988. Physiology of halophilic eubacteria. In: Rodriguez-Valera, F. (Ed.), Halophilic Bacteria. In: Vol. I. CRC Press, Boca Raton, pp. 109–140.

Lennon, J.T., 2007. Diversity and metabolism of marine bacteria cultivated on dissolved DNA. Appl. Environ. Microbiol. 73, 2799–2805.

Lever, M.A., Rogers, K.L., Lloyd, K.G., Overmann, J., Schink, B., Thauer, R.K., Hoehler, T.M., Jørgensen, B.B., 2015. Life under extreme energy limitation: a synthesis of laboratory- and field-based investigations. FEMS Microbiol. Rev. 39, 688–728. https://doi.org/10.1093/femsre/fuv020.

Lipp, J.S., Morono, Y., Inagaki, F., Hinrichs, K.U., 2008. Significant contribution of Archaea to extant biomass inmarine subsurface sediments. Nature 454, 991–994.

Lomstein, B.A., Langerhuus, A.T., D'Hondt, S., Jørgensen, B.B., Spivack, A., 2012. Endospore abundance, microbial growth and necromass turnover in deep sub-seafloor sediment. Nature 484, 101–104.

Mancinelli, R.L., 2000. Accessing the Martian deep subsurface to search for life. Planet Space Sci. 48, 1035–1042.

McEwen, A.S., Ojha, L., Dundas, C.M., Mattson, S.S., Byrne, S., Wray, J.J., Cull, S.C., Murchie, S.L., Thomas, N., Gulick, V.C., 2011. Seasonal flows on warm Martian slopes. Science 333, 740–743.

McGenity, T.J., Gemmell, R.T., Grant, W.D., Stan-Lotter, H., 2000. Origins of halophilic micro-organisms in ancient salt deposits (MiniReview). Environ. Microbiol. 2, 243–250.

McKay, C.P., Stoker, C.R., Glass, B.J., Davé, A.I., Davila, A.F., Heldmann, J.L., Marinova, M.M., Fairen, A.G., Quinn, R.C., Zacny, K.A., Paulsen, G., Smith, P.H., Parro, V., Andersen, D.T., Hecht, M.H., Lacelle, D., Pollard, W.H., 2013. The icebreaker life mission to Mars: a search for biomolecular evidence for life. Astrobiology 13, 334–353. https://doi.org/10.1089/ast.2012.0878.

McKay, C.P., Anbar, A.D., Porco, C., Tsou, P., 2014. Follow the plume: the habitability of Enceladus. Astrobiology 14, 352–355. https://doi.org/10.1089/ast.2014.1158.

Morita, R.Y., 1982. Starvation-survival of heterotrophs in the marine environment. Adv. Microbiol. Ecol. 6, 171–198.

Morita, R.Y., 1997. Bacteria in Oligotrophic Environments. Chapman & Hall, New York, NY.

Muñoz-Iglesias, V., Bonales, L.J., Prieto-Ballesteros, O., 2013. pH and salinity evolution of Europa's brines: Raman spectroscopy study of fractional precipitation at 1 and 300 bar. Astrobiology 13, 693–702.

Nicholson, W.L., Munakata, N., Horneck, G., Melosh, H.J., Setlow, P., 2000. Resistance of *Bacillus* endospores to extreme terrestrial and extraterrestrial environments. Microbiol. Mol. Biol. Rev. 64, 548–572.

Nielsen, H., Ricke, W., 1964. Schwefel-Isotopenverhältnisse von Evaporiten aus Deutschland: Ein Beitrag zur Kenntnis von δ^{34} S im Meerwasser-Sulfat. Geochim. Cosmochim. Acta 28, 577–591.

Norton, C.F., Grant, W.D., 1988. Survival of halobacteria within fluid inclusions in salt crystals. J. Gen. Microbiol. 134, 1365–1373.

Nyström, T., 2004. Stationary-phase physiology. Annu. Rev. Microbiol. 58, 161–181.

Ojha, L., Wilhelm, M.B., Murchie, S.L., McEwen, A.S., Wray, J.J., Hanley, J., Massé, M., Matt Chojnacki, M., 2015. Spectral evidence for hydrated salts in recurring slope lineae on Mars. Nat. Geosci. 8, 829–832. https://doi.org/10.1038/ngeo2546.

Oliver, J.D., Stringer, W.F., 1984. Lipid composition of a psychrophilic marine *Vibrio* sp. during starvation-induced morphogenesis. Appl. Environ. Microbiol. 47, 461–466.

Onstott, T.C., Magnabosco, C., Aubrey, A.D., Burton, A.S., Dworkin, J.P., Elsila, J.E., Grunsfeld, S., Cao, B.H., Hein, J.E., Glavin, D.P., Kieft, T.L., Silver, B.J., Phelps, T.J., van Heerden, E., Opperman, D.J., Bada, J.L., 2014. Does aspartic acid racemization constrain the depth limit of the subsurface biosphere? Geobiology 12, 1–19. https://doi.org/10.1111/gbi.12069.

Onyenwoke, R.U., Brill, J.A., Farahi, K., Wiegel, J., 2004. Sporulation genes in members of the low G+C Gram-type-positive phylogenetic branch (Firmicutes). Arch. Microbiol. 182, 182–192.

Oren, A., 2008. Microbial life at high salt concentrations: phylogenetic and metabolic diversity. Saline Syst. 4. https://doi.org/10.1186/1746-1448-4-2.

Pak, E., Schauberger, O., 1981. Die geologische Datierung der ostalpinen Salzlagerstätten mittels Schwefelisotopenuntersuchungen. Verh. Geol. B-A Jahrg. 1981, 185–192.

Park, J.S., Vreeland, R.H., Cho, B.C., Lowenstein, T.K., Timofeeff, M.N., Rosenzweig, W.D., 2009. Haloarchaeal diversity in 23, 121 and 419 MYA salts. Geobiology 7, 515–523.

Parkes, R.J., Cragg, B.A., Bale, S.J., Getliff, J.M., Goodman, K., Rochelle, P.A., Fry, J.C., Weightman, A.J., Harvey, S.M., 1994. Deep bacterial biosphere in Pacific Ocean sediments. Nature 371, 410–413.

Postberg, F., Kempf, S., Schmidt, J., Beinsen, A., Abel, B., Buck, U., Srama, R., 2009. Sodium salts in E-ring ice grains from an ocean below the surface of Enceladus. Nature 459, 1098–1101.

Radax, C., Gruber, C., Stan-Lotter, H., 2001. Novel haloarchaeal 16S rRNA gene sequences from Alpine Permo-Triassic rock salt. Extremophiles 5, 221–228.

Rampelotto, P.H., 2013. Extremophiles and extreme environments. Life (Basel) 3, 482–485.

Roedder, E., 1984. The fluids in salt. Am. Miner. 69, 413–439.

Rothschild, L.J., Mancinelli, R.L., 2001. Life in extreme environments. Nature 409, 1092–1101.

Rutz, B.A., Kieft, T.L., 2004. Phylogenetic characterization of dwarf archaea and bacteria from a semi-arid soil. Soil Biol. Biochem. 36, 825–833.

Sankaranarayanan, K., Lowenstein, T.K., Timofeeff, M.N., Schubert, B.A., Lum, J.K., 2014. Characterization of ancient DNA supports long-term survival of haloarchaea. Astrobiology 14, 553–560.

Schauberger, O., 1986. Bau und Bildung der Salzlagerstätten des ostalpinen Salinars. Arch. Lagerst forschg Geol. BA 7, 217–254.

Schubert, B.A., Lowenstein, T.K., Timofeeff, M.N., Parker, M.A., 2010. Halophilic archaea cultured from ancient halite, Death Valley, California. Environ. Microbiol. 12, 44–454.

Seckbach, J. (Ed.), 2000. Journey to Diverse Microbial Worlds. Adaptation to Exotic Environments. Series: Cellular Origins and Life in Extreme Habitats. vol. 2. Kluwer, Dordrecht, The Netherlands.

Seckbach, J., Oren, A., Stan-Lotter, H. (Eds.), 2013. Polyextremophiles. Life under multiple form of stress. Series: Cellular Origins, Life in Extreme Habitats and Astrobiology. Springer Dordrecht, Heidelberg/New York/London.

Setlow, P., Kornberg, A., 1970. Biochemical studies of bacterial sporulation and germination. XXII. Energy metabolism in early stages of germination of *Bacillus megaterium* spores. J. Biol. Chem. 245, 3637–3644.

Soppa, J., 2011. Functional genomic and advanced genetic studies reveal novel insights into the metabolism, regulation, and biology of *Haloferax volcanii*. Archaea. https://doi.org/10.1155/2011/602408. Article ID 602408.

Soppa, J., 2013. Evolutionary advantages of polyploidy in halophilic archaea. Biochem. Soc. Trans. 41, 339–343. https://doi.org/10.1042/BST20120315.

Soppa, J., 2014. Polyploidy in archaea and bacteria: about desiccation resistance, giant cell size, long-term survival, enforcement by a eukaryotic host and additional aspects. J. Mol. Microbiol. Biotechnol. 24, 409–419. https://doi.org/10.1159/000368855.

Spötl, C., Hasenhüttl, C., 1998. Thermal history of the evaporitic *Haselgebirge melange* in the northern calcareous alps (Austria). Geol. Rundsch. 87, 449–460.

Stan-Lotter, H., McGenity, T.J., Legat, A., Denner, E.B.M., Glaser, K., Stetter, K.O., Wanner, G., 1999. Very similar strains of *Halococcus salifodinae* are found in geographically separated Permo-Triassic salt deposits. Microbiology 145, 3565–3574.

Stan-Lotter, H., Pfaffenhuemer, M., Legat, A., Busse, H.-J., Radax, C., Gruber, C., 2002. *Halococcus dombrowskii* sp. nov., an Archaeal isolate from a Permian alpine salt deposit. Int. J. Syst. Evol. Microbiol. 52, 1807–1814.

Steen, A.D., Jørgensen, B.B., Lomstein, B.A., 2013. Abiotic racemization kinetics of amino acids in marine sediments. PLoS ONE. 8 (8), e71648, https://doi.org/10.1371/journal.pone.0071648.

Steindler, L., Schwalbach, M.S., Smith, D.P., Chan, F., Giovannoni, S.J., 2011. Energy starved *Candidatus* Pelagibacter ubique substitutes light-mediated ATP production for endogenous carbon respiration. PLoS ONE. 6 (5), e19725.

Stollenwerk, M., Fallgren, C., Lundberg, F., Tegenfeldt, J.O., Montelius, L., Ljungh, A., 1998. Quantitation of bacterial adhesion to polymer surfaces by bioluminescence. Zentralbl. Bakteriol. 287, 7–18.

Thode, H.G., Monster, J., 1965. Sulfur isotope geochemistry of petroleum, evaporites, and ancient seas. Am. Assoc. Pet. Geol. Mem. 4, 367–377.

Treiman, A.H., Gleason, J.D., Bogard, D.D., 2000. The SNC meteorites are from Mars. Planet Space Sci. 48, 1213–1230.

Tsou, P., Brownlee, D.E., McKay, C.P., Anbar, A.D., Yano, H., Altwegg, K., Beegle, L.W., Dissly, R., Strange, N.J., Kanik, I., 2012. LIFE: Life investigation for Enceladus, a sample return mission concept in search for evidence of life. Astrobiology 12, 730–742.

Vreeland, R.H., Jones, J., Monson, A., Rosenzweig, W.D., Lowenstein, T.K., Timofeeff, M., Satterfield, C., Cho, B.C., Park, J.S., Wallace, A., Grant, W.D., 2007. Isolation of live cretaceous (121–112 million years old) halophilic archaea from primary salt crystals. Geomicrobiol J. 24, 275–282.

Webb, K.M., DiRuggiero, J., 2013. Radiation resistance in extremophiles: fending off multiple attacks. In: Seckbach, J., Oren, A., Stan-Lotter, H. (Eds.), Polyextremophiles. Life Under Multiple Forms of Stress. Springer, Dordrecht Heidelberg/New York/London, pp. 251–267.

Whitman, W.B., Coleman, D.C., Wiebe, W.J., 1998. Prokaryotes: the unseen majority. Proc. Natl. Acad. Sci. U. S. A. 95, 6578–6583.

Zacny, K., Paulsen, G., McKay, C.P., Glass, B., Davé, A., Davila, A.F., Marinova, M., Mellerowicz, B., Heldmann, J., Stoker, C., Cabrol, N., Hedlund, M., Craft, J., 2013. Reaching 1 m deep on Mars: the icebreaker drill. Astrobiology 13, 1166–1198. https://doi.org/10.1089/ast.2013.1038.

Zerulla, K., Soppa, J., 2014. Polyploidy in haloarchaea: advantages for growth and survival. Front. Microbiol. 5. https://doi.org/10.3389/fmicb.2014.00274.

Zerulla, K., Chimileski, S., Näther, D., Gophna, U., Papke, R.T., Soppa, J., 2014. DNA as a phosphate storage polymer and the alternative advantages of polyploidy for growth or survival. PLoS ONE. 9, e94819.

Zharkov, M.A., 1981. History of Paleozoic Salt Accumulation. Springer Verlag Berlin.

Zolensky, M.E., Bodnar, R.J., Gibson, E.K., Nyquist, L.E., Reese, Y., Shih, C.Y., Wiesman, H., 1999. Asteroidal water within fluid inclusion-bearing halite in an H5 chondrite, Monahans (1998). Science 285, 1377–1379.

Under what kind of life system could space life emerge?

9

Kenji Ikehara

G&L Kyosei Institute, Nara, Japan The International Institute for Advanced Studies of Japan, Kyoto, Japan

Chapter outline

1 Introduction

Astrobiology is a research field studying on what kind of space organisms (S-organisms) are living under what kind of life system or genetic system and on consequences which an Earth organism (E-organism) is suffered from environments in space when the organism is placed in there, and so on. In fact, explorations of planets in Solar system, as Mars, and their satellites of Jupiter and Saturn, as Europa, Enceladus, and Titan, have been carried out by NASA to find out some traces of life. In

Japan, it was widely informed through newspapers, and so on, at the time when Japanese spacecrafts, Hayabusa and Hayabusa 2, which were designed to bring asteroid samples back to Earth, launched to asteroids, "Itokawa" and "Kaguya," that a clue for solving the origin of life might be brought.

On the other hand, human beings know only organisms living on one planet, Earth, at this time point. So, it will become possible to understand whether the life system, under which E-organisms are living, is unique or universal, if the second S-organism and third one could be discovered and human beings could understand the life system of the organisms how they emerged and live on the respective planets or satellites. Therefore astrobiology is one of important and quite interesting scientific fields for not only exploration of space life (S-life) and but also for basic science of Earth life (E-life).

It is also well known that many kinds of organisms inhabit in extreme environments on Earth, such as high and cold temperature, highly acidic and basic water, and so on. Many researchers frequently consider that the study on the extremophiles might lead to discovery of S-organisms, because it is expected that such extreme environments on Earth should be also present somewhere in extraordinary wide space. However, it is totally unknown about chemical materials with which S-organisms are living and whether organisms, which emerged on a planet in space, have evolved similarly to the organisms living on the Earth, or not. S-life might even extinct just after the emergence of the life because of an abrupt climate change and collision of a small planet, and so on, so that the dead organisms are fixed.

Of course, it would be generally important to grasp objects studied widely as much as possible, when unknown objects such as S-organisms are studied. Otherwise, it might be failed to notice a fact due to a shortsighted mind. However, it would be made difficult to obtain effective results as the consequence to grasp the object too widely and to reach meaningful conclusion in spite of many strenuous efforts and waste of much research expense because of easygoing investigations.

We, human beings, know one important object for us to consider about organisms in space. In addition, we understand the object in quite detail, for example, about how the organisms are living on one planet under what kind of life system (Gargaud et al., 2012; Ikehara, 2002, 2005, 2016a, b). Those are just E-organisms, which emerged and evolved on one small but wonderful water-planet. Therefore it must be considered deeply about what kind of possibility exists if organisms emerged and are living on a planet or a satellite somewhere in space, through both investigating on the mechanism under which E-organisms are living and considering about the reason how E-organisms could obtain the mechanism. The study on the origin of E-life should be the most fundamental one for extrapolation of space organisms and should be one of the most important research fields of astrobiology. So, the first step of the life-origin study in astrobiology must be to make clear how the first life emerged on the primitive Earth. Thus I would like to assert that the astrobiological study should be taken more attention to the origin of E-life than before. Then, exploration of life on a planet or satellite in space should be practiced as the second step of the study on S-life under background of the origin of life on the Earth.

Thus, first, it is explained according to [GADV]-protein world hypothesis (GADV hypothesis) in the next Section, how E-organisms emerged on this planet about 3.8 billion years ago and evolved to modern organisms (Ikehara, 2002, 2005, 2016a, b). Capital letters, as GADV, in square brackets indicate amino acid written by one letter symbol to discriminate amino acids, Gly [G] and Ala [A], especially from nucleo-bases, guanine G and adenine A. So, [GADV] means four amino acids [G], [A], Asp [D], and Val [V]. Square brackets are omitted only in the case of the special term, GADV hypothesis, because it is clear that GADV, even without the brackets, means the four amino acids. Thereafter, it will be discussed the way how S-organism could emerge and evolve under what kind of life system and then about under what kind of life system space organisms are living, if some S-organisms are actually inhabiting somewhere on planets or satellites in space.

2 How did life emerge on the primitive Earth?

Many hypotheses have been proposed to solve the riddle on the origin of life, how E-life emerged on the primitive Earth such as Panspermia hypothesis, space-origin hypothesis, hydrothermal vent hypothesis (Imai et al., 1999; Holm and Andersson, 2005; Chandru et al., 2013), RNA world hypothesis (Gilbert, 1986), coenzyme world hypothesis (Sharov, 2016), amyloid world hypothesis (Maury, 2015) and tRNA core hypothesis (De Farias et al., 2016), and so on. However, the evolutionary process from accumulation of organic compounds on the primitive Earth to the emergence cannot be rationally explained by any hypotheses including RNA world hypothesis (Ikehara, 2017).

On the other hand, I have proposed [GADV]-protein world hypothesis (GADV hypothesis) on the origin of life about 15 years ago (Ikehara, 2002), based on the analysis of databases of genes and proteins with the four conditions (hydrophobicity/hydrophilicity, α-helix, β-sheet, and turn/coil formabilities) for protein structure formation, which were obtained with data of extant microbial genes encoding water-soluble globular proteins (Ikehara et al., 2002; Ikehara, 2002). The hypothesis assumes that life emerged from [GADV]-protein world, which was formed by pseudo-replication of [GADV]-protein in the absence of any genetic function (Ikehara, 2009). I have asserted that the evolutionary process to the emergence of life can be explained according to the GADV hypothesis without any large contradiction. The GADV hypothesis, but not RNA world hypothesis, can also well explain three formation processes of the first protein, genetic code, and gene (Ikehara, 2017). On the contrary, the processes have not been addressed in other hypotheses, such as cofactor world hypothesis, amyloid world hypothesis, tRNA core hypothesis, and so on.

Here, I describe an evolutionary process from synthesis of organic compounds including amino acids on the primitive Earth to the emergence of life in more detail (Ikehara, 2002, 2005, 2016a, b): (1) Primitive atmosphere composed of CO_2, H_2, H_2O, N_2, CH_4, NH_3, and so on was formed. (2) Simple amino acids, as

[GADV]-amino acids, were physically and chemically synthesized and accumulated on the primitive Earth. (3) Peptide catalysts mainly composed of [GADV]-amino acids were produced such as by repeated heat-drying process in depressions on the primitive Earth. (4) [GADV]-protein world was formed by pseudo-replication with [GADV]-protein(s), actually [GADV]-peptide aggregate(s). (5) Nucleotides and oligonucleotides were synthesized and accumulated in the protein world. (6) The first GNC genetic code encoding [GADV]-amino acids was established through complex formation of GNC-containing oligonucleotide or primeval tRNA with the corresponding [GADV]-amino acid. (7) Single-stranded $(GNC)_n$ genes were produced by joining of GNC triplet base sequence in the complex. (8) Successively, double-stranded $(GNC)_n$ genes were formed by synthesis of complementary strand to the single-stranded $(GNC)_n$ genes. (9) Finally, the first life emerged on the primitive Earth, after a sufficient number of double-stranded $(GNC)_n$ genes, which are required for the first life to emerge, were produced.

The reason, why the establishment processes of the three elements, the first gene, the first genetic code, and the first protein, could be explained with GADV hypothesis for the first time, would be because I introduced a new concept or protein 0th-order structure (Ikehara, 2002, 2005, 2014). Therefore it is expected that life emerged from [GADV]-protein world, but not from RNA world (Ikehara, 2017). Then, the emergence and evolution of E-life are discussed according to GADV hypothesis in the next section.

3 Why could Earth organisms emerge from [GADV]-protein world?

In this Section, it is discussed based on GADV hypothesis, the reason why life could emerge on the primitive Earth. Consideration of the reason is quite important, because it might be possible to show that only one evolutionary process to emergence of life exists, even if life could emerge on any planets and satellites in space. For the purpose, consider first characteristics of amino acids or [GADV]-amino acids, which led to the emergence of life.

3.1 Characteristics of amino acids

3.1.1 α-Amino acid: Small but wonderful organic molecule

At first, properties of [GADV]-amino acids are described here in order to clarify roles played by [GADV]-amino acids in the emergence of Earth life (Ikehara, 2016a), since general properties of amino acids are described in every text book of Biochemistry, including Stryer's text book, but peculiarities of [GADV]-amino acids are not specially described (Stryer, 1988). [GADV]-amino acids are only four amino acids in extraordinary large number of amino acids. Common part of the amino acids are composed of only nine atoms (2C, 4H, 2O, and 1N) (Fig. 1), all of which belong to the first and the second periods of atom's periodic table, meaning

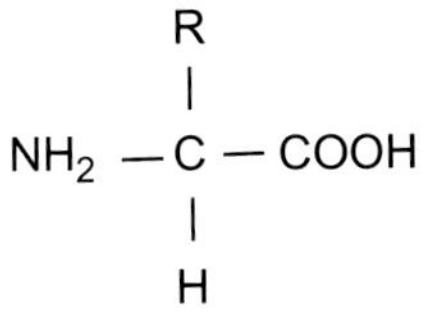

$$
\begin{array}{c}
R \\
| \\
NH_2 \ —\ C\ —\ COOH \\
| \\
H
\end{array}
$$

FIG. 1

Structure formula of amino acid. R is generalized symbol of a side chain. Amino acid is a simple organic compound, which can be regarded as a complex of methane (CH_4), ammonia (NH_3), Formic acid (HCOOH) and alkyl residue (R).

Table 1 Summarized properties of amino acid, peptide bond, [GADV]-amino acids, and [GADV]-polypeptide.

Amino acid

1. Atom: H, C, N, O, (S) (all atoms belong to 1st and 2nd periods, except for S-containing amino acids, cysteine and methionine)
2. The number of atoms in common part: NH_2—(C—)H—COOH = 9
3. Formula weight: Common part (NH_2—(C—)H—COOH = 74)
4. Components of common part
CH_4 (C—(H_4)) + NH_3 (NH_2—(H)) + HCOOH ((H)—COOH)
5. Positive charge: —NH_2 (—NH_3^+), negative charge: —COOH (—COO^-)

Peptide bond

1. Half double bond (Restriction of free rotation around N—C bond)
2. Hydrogen donor (—NH—); Hydrogen accepter (—CO—)

[GADV]-amino acids

1. Atom: H, C, N, O (all atoms belong to 1st and 2nd periods)
2. The number of atoms: Gly = 10, Ala = 13, Asp = 16, Val = 19
3. Molecular weight: Gly = 75, Ala = 89, Asp = 133, Val = 117
4. Homochiral (Three amino acids (Ala, Asp, Val) are L-form except for glycine without asymmetric carbon atom)

[GADV]-polypeptide

1. Secondary structure formability
α-helix (Ala), β-sheet (Val), turn/coil (Gly (Asp))
2. Tertiary structure formability
Hydrophilic surface (Asp (Gly)), Hydrophobic core (Val (Ala))

that the atoms themselves are small. The common part of α-amino acids can be regarded as a complex formed with quite simple two organic and one inorganic compounds, methane (CH_4 = 5), formic acid (HCOOH = 5), and ammonia (NH_3 = 4) (Fig. 1, Table 1). Numeral written in parentheses is the total number of atoms of the molecule.

Amino acids also have a side chain, which is different from each other and expresses characteristics of the respective amino acids. The total number of atoms contained in [GADV]-amino acids is quite small as Gly = 10, Ala = 13, Asp = 16, Val = 19 (Table 1). Then, [GADV]-amino acids can be easily produced with prebiotic means, such as electric discharge into primitive atmosphere (Miller and Orgel, 1974). Irrespective of the smallness of the organic molecules, [GADV]-amino acids have many splendid physical and chemical properties and are core molecules, which could lead to the emergence of Earth life.

All of four [GADV]-amino acids have α-carbon atom, with which both amino and carboxyl groups in addition to one hydrogen are directly bound. The α-carbon atom also directly connects with a side chain, which differentiates from other amino acids. Thus polypeptide chain produced by joining of [GADV]-amino acids can form regular structures, such as α-helix and β-sheet, owing to the common chemical structure.

3.1.2 Chirality of natural amino acids used in Earth organisms

As well known, all amino acids utilized in extant organisms on this planet are L-amino acids or L-α-amino acids, except for Gly, which has not asymmetric carbon atom. The restriction of the amino acids into L-form assists formation of regular structures, such as α-helix and β-sheet structures, because the side chain of amino acids tends toward the same direction, so that steric hindrance among side chains can be effectively avoided.

3.1.3 Amino acids have both positive and negative charges in the molecules

[GADV]-amino acids carry positive charge on amino group ($-NH_3^+$) and negative charge on carboxyl group ($-COO^-$) in neutral water (Fig. 2). Therefore two amino acids could pull against each other through electrostatic interaction in water containing various kinds of organic compounds, which were accumulated on the primitive Earth. The specific association could induce to efficiently form peptide bond between two amino acids at a high probability to exclude various organic compounds other than amino acids from polypeptide.

3.1.4 Partial double bond character of peptide bond

Peptide bond formed between two amino acids exhibits partial double bond character because of resonance between two canonical structures (Fig. 3A). The double bond character causes to restrict free rotation around N—C bond in the peptide bond.

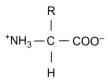

FIG. 2

Structure formula of amino acid ionized at neutral pH in water. Amino acid has both positive charge ($-NH_3^+$) and negative charge ($-COO^-$) in the small molecule.

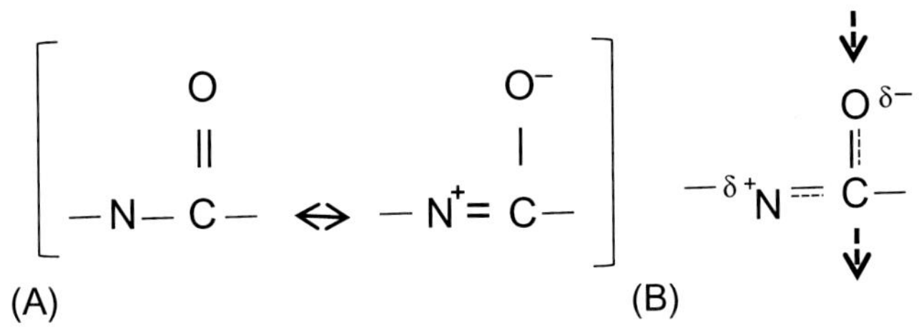

FIG. 3

(A) Resonance between two canonical formulas of peptide bond. (B) Because of the resonance structures, the peptide bond exhibits partial double-bond character, which prevents free rotation about N—C bond. Downward and upward broken arrows indicate proton donor and accepter, respectively.

Therefore the number of free rotatable single bonds is restricted into two per one amino acid residue in the main chain of polypeptide composed of α-amino acids. As a result, excess free rotation in the polypeptide is repressed and sufficient free rotation is simultaneously guaranteed. This makes it possible to form easily stable secondary structures, as right-handed α-helix and β-sheet structures.

Accompanied by the resonance structure of peptide bond, imino and carbonyl groups in the bond become good hydrogen donor and hydrogen accepter, respectively, so that polypeptide chain can form stable secondary structures through formation of the hydrogen bonds in a main chain or between two chains, such as α-helix and β-sheet structures (Fig. 3B).

3.1.5 Expression of protein function

Polypeptide, which is formed by polymerization of amino acids, is a single-stranded chain. The order of amino acid sequence of extant proteins is determined by genetic information written as base sequence of DNA. The resulting polypeptide chain forms mainly three secondary structures such as α-helix, β-sheet, and turn/coil structures through various short-range interactions, which depend on side chain of amino acid residues. Successively, the resulted polypeptide chain is folded into tertiary structure, usually into water-soluble globular structure, through short- and long-range interactions among amino acid residues. Consequently, amino acid residues, which are separated on the amino acid sequence, frequently become close to each other on tertiary structure. The amino acid residues, which come close to each other depending on necessity, form functional site as catalytic center for chemical reaction. The catalytic activity is controlled through structure change called as allosteric transition, which is triggered by binding of a chemical compound at a site separated far from the catalytic site. Such a structure change which is induced by a chemical compound bound at one end of rod-like α-helix can be transmitted to the other end of the helix. Structure change of a protein is also induced by generation or degradation of a part of the secondary structures.

Such kind of skillful structure change inducing control of chemical reaction looks like just as polymeric molecular precision machine. Features of amino acids make it possible to form such a wonderful molecular machine. Contrary to that, single-stranded DNA and RNA cannot generally form regular structure as helix. This indicates that amino acids have many splendid properties for formation of functional proteins.

3.2 [GADV]-amino acids played the important roles in the emergence of Earth life

3.2.1 Characteristics of [GADV]-amino acids for formation of secondary and tertiary structures

The roles of [GADV]-amino acids in formation of three secondary structures (α-helix, β-sheet, and turn/coil) are successfully divided into the respective three amino acids out of four. That is, as seen in Table 1, Gly, Ala, and Val are turn/coil, α-helix, and β-sheet forming amino acids, respectively. The division of the roles among [GADV]-amino acids makes it possible to form appropriately accented water-soluble globular protein with three secondary structures. For example, segments collecting of Gly, Ala, and Val at a high frequency exhibit a high propensity for forming turn/coil, α-helix, and β-sheet, respectively.

Asp and Val are only one highly hydrophilic and hydrophobic amino acids in [GADV]-amino acids, respectively. Therefore segments containing Asp and Val at a high ratio locate on surface and in core regions of water-soluble globular [GADV]-protein or aggregate of [GADV]-peptides. Thus [GADV]-protein can show a large stability in water, because both strongly hydrophilic and hydrophobic amino acids are also successfully contained in only four [GADV]-amino acids.

The four, hydrophilicity/hydrophobicity, α-helix, β-sheet, and turn/coil formabilities of a protein, were obtained as average values of protein structure indexes, which are calculated with multiplication of an amino acid index with the corresponding amino acid composition of an extant protein encoded by genome of seven microorganisms carrying the, respectively, different GC contents (Ikehara et al., 2002). Therefore there generally exist two ways on how four amino acids satisfy the four conditions for formation of water-soluble globular protein. One is the case where the four conditions are satisfied by four amino acids with medium propensity for secondary and tertiary structure formation. In this case, every segment in a protein should exhibit similarly featureless secondary and tertiary formabilities. Another is the case where the conditions are satisfied by four highly characteristic amino acids with high or low secondary and tertiary structure propensities. In this case, the respective segments could fulfill featured properties for secondary and tertiary structure formation, so that many segments in a protein could be folded into accented secondary and tertiary structures. [GADV]-protein and extant natural protein are the latter case.

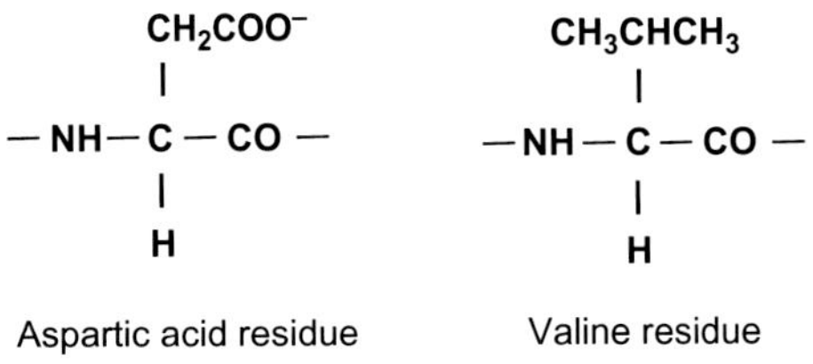

$$CH_2COO^-$$
|
$$— NH—C—CO —$$
|
H

Aspartic acid residue

$$CH_3CHCH_3$$
|
$$—NH—C—CO —$$
|
H

Valine residue

FIG. 4

Structure formula of aspartic acid (A) and valine residues (B). The strong hydrophilic (Asp) and hydrophobic (Val) residues make it possible to fold [GADV]-polypeptide chain into water-soluble globular structure in water.

3.2.2 The role of Val in formation of water-soluble globular protein structure

Gly having hydrogen atom at the side chain is the most simple and is the most easily synthesized with prebiotic means among the four [GADV]-amino acids (Miller and Orgel, 1974). Contrary to that, Val with isopropyl group as the side chain is the most complex amino acid and the most difficult to synthesize with prebiotic means in the four amino acids (Fig. 4). Therefore pools containing roughly equal amounts of [GADV]-amino acids did not actually exist on the primitive Earth, because Val would accumulate at the lowest concentration among [GADV]-amino acids on the primitive Earth. However, peptides containing Val could be preferentially assembled into aggregate of [GADV]-peptides through hydrophobic interaction in water. Consequently, imbalance among the concentrations of [GADV]-amino acids accumulated on the primitive Earth could be relatively corrected to form water-soluble globular protein at a high probability.

3.2.3 [GADV]-amino acids is the simplest combination for formation of water-soluble globular protein

The number of atoms of [GADV]-amino acids, Gly, Ala, Asp, and Val, are 10, 13, 16, and 19, respectively (Table 1). The numbers of atoms in the side chain of the respective amino acids are 1, 4, 7, and 10. Gly and Ala are the first and the second smallest amino acids among 20 natural amino acids, which are arranged in order from the smallest number of atoms in the side chain. This means that Gly and Ala are the simplest turn/coil and α-helix forming amino acids, respectively. Furthermore, Asp and Val are the simplest hydrophilic and hydrophobic amino acids, respectively. In addition to that, Val is the smallest β-sheet forming amino acid among 20 natural amino acids. These indicate that [GADV]-amino acids is the simplest combination among all four amino acids, with which the four conditions for formation of water-soluble globular protein can be satisfied. This was confirmed by the fact that every combination of four amino acids, which was obtained by replacement of one of [GADV]-amino acids with an amino acid simpler than the replaced amino acid, did not satisfy the four conditions (Ikehara, 2002).

2-Aminobutylate (2-AB) is α-amino acid with ethyl group on a side chain (the number of atoms in ethyl group is 7). Therefore 2-AB is a similarly small to aspartic acid and a simpler amino acid than valine. Therefore 2-AB is easily synthesized with prebiotic means and could accumulate on the primitive Earth (Miller and Orgel, 1974). Nevertheless, 2-AB was not used in the primeval GNC genetic code and is not contained in 20 natural amino acids. The reason can be explained as follows. 2-AB is α-helix forming amino acid, because 2-AB does not carry a branched or bulky side chain at the β-carbon atom similarly as α-helix forming amino acids, such as Ala, Glu, Leu, Met, and so on. On the other hand, Ala with methyl group at a side chain could more easily accumulate on the primitive Earth than 2-aminobutylate. Therefore not 2-AB but Ala was used as α-helix forming amino acid in GNC primeval genetic code. This clearly means that the four [GADV]-amino acids are selected out, not only because the amino acids are small and simple but also because water-soluble globular protein composed of [GADV]-amino acids has sufficiently high secondary and tertiary formabilities.

3.2.4 Why could life emerged from [GADV]-protein world on the primitive Earth?

Three problems, which are closely entangled to each other, have mainly made it quite difficult to solve the riddle on the origin of life.

The first problem is to explain the formation process of "chicken and egg relationship" between gene and protein. Although many persons might consider that RNA world hypothesis already resolved the problem of "chicken and egg relationship," since the hypothesis was proposed to resolve it (Gilbert, 1986). However, the formation process can be also explained according to the GADV hypothesis, probably more reasonably than the RNA hypothesis, as that the relationship was formed as going up from the lower ([GADV]-protein world) to the upper stream (gene) of the genetic flow in the present life system (Ikehara, 2002, 2005, 2016b).

The second one is to explain conversion process from random to systematic polymer synthesis, because it is generally supposed that only meaningless polypeptide and RNA should be produced by the random polymerization of the respective monomeric units on the primitive Earth. Contrary to that, many polymers with ordered sequence are produced in modern organisms under the genetic function. Therefore it would be essential for elucidating the riddle on the origin of life to make clear the process how systematic polymer synthesis became possible during random processes. I consider that protein 0th-order structure or [GADV]-amino acids, in which random joining could produce water-soluble globular protein in the absence of genetic function, played the key role in converting from random to ordered polymerization on the primitive Earth (Ikehara, 2014, 2017).

Third one is to account for a process how the first genetic system composed of gene, genetic code, and protein could be established on the primitive Earth. In this case too, the establishment process of the first genetic system can be explained based on GADV hypothesis without any large contradiction, as that the system was created as proceeding from globular aggregates of short GADV-rich peptides to more complete water-soluble globular [GADV]-proteins step by step (Ikehara, 2002, 2005, 2016b).

As described before, the first life emerged after formation of [GADV]-protein world, the first GNC genetic code and double-stranded $(GNC)_n$ gene, owing to splendid characteristics of [GADV]-amino acids and the protein composed of the four amino acids.

3.3 Under what kind of life system can entirely new life reemerge on the Earth?

Next, let us consider what kind of life with a high ability similarly to the first *E*-life can emerge again on a future Earth under what kind of life system, after all organisms exterminate and all organic compounds disappear from the Earth upon an imaginary incident. The disappearance of all organic compounds from the Earth should be required for the entirely new life to reemerge on the Earth, because complex organic compounds, which were used in the previous organisms and remained on the Earth, should cause obstacles for the new life to emerge. Of course, although I do not want even to imagine such a situation, the answer to such a question could provide a clue for solving a similar problem about what kind of life could emerge on a planet somewhere in space other than the Earth.

Can entirely new life reappear on a water-planet, Earth, after what kind of steps will progress from the extermination of E-life. The processes, under which the first life emerged on the primitive Earth should be referred and the following criteria should be taken into consideration, in order to avoid desultory discussion.

(1) Procreative power of the second life, which might reappear under a fundamental life system on an imaginary future Earth, is at least similarly equal to or higher than the life system of the first life. The reason, why this criterion must be set, is because it should be expected for a newly reappearing organism to meet with many difficulties similar to those encountered when the first life emerged on the primitive Earth. Otherwise, new life could never reemerge on the future Earth.

(2) The second life should also live with organic catalysts in water environments on the future Earth, too, because the future Earth would be also water-planet.

(3) Structure of monomeric units composing the organic catalysts should be similarly simple to [GADV]-amino acids, because the monomeric units must accumulate without using genetic system for the new life to reemerge.

(4) Organic catalysts of the second life must have secondary and tertiary structures similarly to proteinaceous catalysts used in the first life, because the catalysts must exhibit its activity like a molecular precision machine. Otherwise, the second life with a high ability could never reemerge on the future Earth.

(5) Monomeric units composing the organic catalysts should lead to formation of genetic system, with which the second life can be reproduced. Otherwise, the second life will also never reemerge on the future Earth.

As described in Section 3.2.3, [GADV]-amino acids are the simplest and the most suitable combination, in which all factors necessary for formation of water-soluble globular protein are admirably contained. Therefore I believe that the second life also will emerge from [GADV]-protein world, because it is easily anticipated that there is no route for the emergence of life other than that expected from GADV hypothesis.

However, it must also take notice of the following matter. It is questionable to insist that there is no other way than that assumed by GADV hypothesis to reach to the emergence of life from simple organic compounds, even if evolutionary steps to the emergence of life could be rationally explained based on the GADV hypothesis. The reason is because a rationale matter is not always correct and facts beyond a limit supposed by researchers have been frequently discovered later. Therefore it must be taken notice that there is a limit of wishful thinking. However, I would like to assert repeatedly that it is important to propose various reasonable ideas and to discuss them impartially as a scientist, because such attitude could stimulate serious discussion on the origins of S-life and E-life.

4 How can organisms emerge on planets or satellites in space?

Diverse physical and chemical environments should exist on surface of planets and satellites in space, for example, extremely low and high temperature, pond or sea of hydrocarbon or alcohol, various kinds of atmospheres composed of a high concentration of carbon dioxide and methane, and so on. Furthermore, even the environments on the planets and the satellites, which circulate on a constant orbit around a fixed star would change since light energy emitted from the star should change as the time lapses.

Then, let us consider about the theme what kind of life could emerge on a planet or satellite in space, under what kind of life system in the following sections.

4.1 Environment for the emergence of life in space

Environmental situation of planet surface is mainly determined by physical conditions of the planet itself, such as a distance from a mother star, size of the planet and age of the mother star and the planet, and so on (Gargaud et al., 2012). Therefore there exist diverse environments on the surface of planets in space. This means that it might be possible for space organisms composed of various kinds of chemical compounds to emerge. As described before, the most essential matter for life to emerge should be organic catalysts and a genetic system, with which the catalysts can be reproduced. Then, the possibilities that space organisms emerge and evolve using what kinds of organic catalysts are discussed in the following sections.

4.1.1 Can space life emerge in what kind of phase, gas, liquid, or solid?

Materials can be generally present in three phases, gas, liquid, and solid. Can organisms emerge in which phase out of the three phases? Generally speaking, life could not emerge in solid phase, because movement of chemical compounds is severely restricted in solid phase and, therefore, chemical reaction cannot proceed at a high rate in the phase. Then, can organisms emerge in gas phase? It would be also difficult to promote chemical reactions efficiently, because concentration of chemical compounds is too low to stimulate chemical reactions in the phase at a sufficiently high rate for life to emerge, although chemical reactions in gas phase are well known. As a matter of course, all extant organisms are living under chemical reactions, which are carried out in cells full of water. Therefore ancestors of insects and birds flying in the air also emerged in water, probably in primitive sea and pond, and advanced into the air as living field at some time point as a consequence of the evolution. Of course, chemical compounds could be produced in gas phase or in the primitive atmosphere and accumulate on solid phase or land or in liquid phase, sea or pond. These mean that the first life did not emerge in gas phase or at least it would be quite difficult to emerge directly in the atmosphere.

4.1.2 Can space life emerge in what kind of solvent, water, or other organic solvents?

Next, consider in what kind of solvent S-organisms can emerge. It is well known that extraordinary diversified conditions are observed in space. For example, it is confirmed with *Cassini* and *Huygens* probes that a large quantity of hydrocarbons, as methane and ethane, exists on Saturn VI satellite, Titan, and it forms hydrocarbon seas and lakes (Raulin et al., 2012). Some scientists expect that life might emerge in such hydrocarbon seas and lakes. Furthermore, the spacecraft, *Cassini*, has also found that Saturn II satellite, Enceladus, harbors a global ocean of salty water under its icy crust, possibly with hydrothermal vents on its seafloor (Barge and White, 2017). It is confirmed that the surface of Jupiter II satellite, Europa, is frozen, covered with a layer of ice, but it is also expected that there is an ocean beneath the ice crust. Therefore it would be adequate to consider generally in what kind of solvent life can emerge on planets and satellites in space. The reason is because it is difficult to discuss all of the individual and specific situations and it would be sufficient to understand based on the general discussion under what kind of situations in space life can emerge.

Liquid solvents are largely classified into three: hydrophobic or apolar solvent such as hydrocarbon, intermediately hydrophobic or hydrophilic solvent such as alcohol, and polar solvent such as water. First, consider whether life can emerge in hydrophobic solvent or not. In such a solvent, surface of globular organic catalyst must be hydrophobic in order to interact strongly between hydrophobic solvent and the surface of the catalyst so that the organic catalyst can be stably located in the

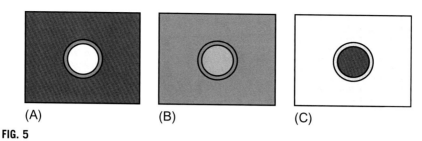

FIG. 5

Schematic drawing of globular polymer solubilized in apolar (A), intermediately hydrophobic (B), or polar (C) solvent. More whitish color indicates that polarity of solvent, surface and inner core of the globular polymer is larger.

solvent (Fig. 5A). In this case, the organic catalyst with hydrophobic surface could not catalyze chemical reaction smoothly, because hydrophobic surface residue cannot strongly pull out or give electron on atom or between two atoms to cut off a chemical bond or to newly form the bond.

Then, what kind of phenomenon can be observed in polar solvent, such as water? Surface of organic catalyst should be hydrophilic to strongly bind with solvent molecules with electrostatic interaction (Fig. 5C). The ability as a catalyst would become sufficiently high, because hydrophilic or polar surface of the catalyst can easily pull out electron from and supply to chemical bond to induce scission or formation of chemical bond.

Further, consider about chemical reaction, which are carried out with organic catalyst in intermediate hydrophobic solvent, for example, alcohol. Interaction between solvent molecule and surface of the catalyst should be also become intermediate, half hydrophobic and half hydrophilic. Such an organic catalyst should exhibit intermediately strong or weak activity in such an intermediately hydrophobic solvent (Fig. 5B). Stability of the catalyst would be not large and not small, because difference of hydrophobicity between surface residue and inner core becomes small. Otherwise, the catalyst with strong hydrophobic inner core should be destroyed by hydrophobic interaction with hydrophobic group of solvent molecules and also the structure of the catalyst with large polarity in the inner core would be also degraded with electrostatic interaction between hydrophilic core and hydrophilic group of solvent molecules. This means that catalytic activity of the catalyst with intermediately hydrophilic surface should inevitably become low.

4.1.3 What kind of life could emerge in space, if the life emerged in water?

As discussed in the previous sections, life would never emerge in a gas and a solid phase, no matter how various and extremely unique environments would exist in space. Therefore S-life should emerge in a liquid phase, if the life could emerge somewhere in space. If so, the life must emerge in a polar solvent such as water but not in organic solvent such as hydrocarbon, alcohol, and so on. The reason is because it is easy to scissor a chemical bond or to form a new chemical bond between two atoms due to the high polarity of the surface of the globular polymer and the polar solvent.

4.2 Can space-life emerge under what kind of life system in water?

What kind of life could emerge under what kind of life system or genetic system on a planet or a satellite in space, if S-life could emerge in water environment?

4.2.1 What is the minimum condition for the emergence of space-life?

If S-life could emerge in water environment, similar or the same conditions with those for reappearance of E-life could be applied to the emergence of S-life, because physical and chemical conditions are the same everywhere irrespective of Earth or other planets and satellites in space. Therefore consider about under what kind of life system S-life could emerge in water.

(1) Procreative power of S-life should be at least similarly equal to or higher than the life system of the first life, which emerged on the primitive Earth.
(2) The S-life should also emerge with organic catalysts acting in water, because chemical reactions are indispensable for the life to emerge.
(3) Monomeric units composing the organic catalysts should be similarly simple or simpler than [GADV]-amino acids, because organic catalysts must be produced with prebiotic means.
(4) Organic catalyst of the S-life should be made with polymer forming secondary and tertiary structures similar to protein to behave like a molecular precision machine and to express a high catalytic activity.
(5) Genetic system or life system, which is essential to the emergence of life, must be also established in relation to the monomeric units composing the organic catalysts.

Only [GADV]-amino acids and [GADV]-protein could satisfy the conditions or criteria for organic catalyst for the emergence of life in water environments, as described in Section 3.2.4. Furthermore, there would not exist any combination of organic compounds with solvent, which is superior to that of [GADV]-protein/ [GADV]-amino acids and water. However, even many researchers working in the field of astrobiology might intuitively deny the idea. Then, I would like to ask them "can you show me simple organic compounds, which satisfy the above conditions?." The behaviors of amino acids or [GADV]-amino acids in water are excellent to lead the emergence of life and proteins composed of the simple amino acids can play skillful roles in stimulating many chemical reactions, as described in Section 3. Therefore I believe that there exists no organic compound except for [GADV]-protein and its components, [GADV]-amino acids, which can satisfy the previous conditions. Thus it can be concluded here that S-life would emerge with [GADV]-amino acids, if the life could emerge somewhere in space similarly as the first life, which emerged on the primitive Earth. I cannot even imagine the other way for life to emerge in water, although the possibility would not be zero, that S-life emerged or will emerge under a life system, which is quite different from that assumed by GADV hypothesis.

4.3 Significance of astrobiology in studies on the origin of life

Sometimes, a possibility, that S-life using silicon or arsenic instead of carbon or phosphorous inhabits somewhere in space, is discussed in the field of Astrobiology. The reason is because carbon and silicon or phosphorous and arsenic belong to the same family of atoms and have similar chemical properties. However, I would like to clearly deny such a possibility because carbon (atomic number $= 6$) and phosphorous (atomic number $= 15$) are much simpler atoms than silicon (atomic number $= 14$) and arsenic (atomic number $= 33$), respectively. Therefore carbon and phosphorous should be much more easily formed and accumulated at a larger quantity on the primitive Earth than silicon and arsenic, respectively. It is also unquestionable that carbon and phosphorous are much superior atoms, which are useful for constructing biomolecules, than silicon and arsenic. These mean that every S-life should use not silicon and arsenic but carbon and phosphorous. In addition, as described in the previous Paragraph, I also firmly believe that every S-life would emerge with [GADV]-amino acids and [GADV]-protein, if life could emerge somewhere in space. Thus I consider that S-life using carbon and phosphorous should emerge according to GADV hypothesis, although figures and shapes of the S-life might be quite different from our E-life because of largely different environments between planet or satellite in space and this planet, Earth.

However, Astrobiology is not always insignificant in the life-origin studies, even if S-life has emerged according to the processes assumed by GADV hypothesis. No, it is not the case. Significance of Astrobiology should inversely increase, because it can be confirmed only by Astrobiology, whether S-life did really emerge according to the hypothesis or not. Furthermore, it has been a dream of human beings for many years to know that life has emerged somewhere in space. We would not always expect that S-life has emerged under a mechanism different from that of E-life. We, human beings, would feel stronger affinities with S-life, if the life has emerged under the fundamental life system, composed of gene, genetic code, and protein, which is similar to E-life.

Of course, I will be totally careless about that, even if S-life did not emerge according to the processes expected by GADV hypothesis, which I have proposed. If GADV hypothesis were wrong, it is the time like that I know the reason why I misunderstood the evolutionary processes to emergence of life, and that I can have a broader life view than before. Contrary to that, it would be a serious problem to consider vaguely that S-life, which uses life system different from ours, could emerge under various environmental conditions in space. When a thing is considered more deeply, probability, that the anticipation comes true, should become larger and countermeasures against the wrong anticipation could be hit upon soon.

4.4 What should be paid attention to, when astrobiologists search for life in space?

As described before, investigation of S-life is undoubtedly quite important. The reason is because it is important to confirm whether S-life has actually emerged on a planet or satellite other than the Earth, or whether the S-life had emerged according to GADV hypothesis, or not. However, I would like to stress here that space investigation should be carried out what should be investigated after it is considered deeply, because investigation without focusing on a research object may overlook an important but faint indication of life in space. Therefore space investigation should be carried out toward the object specified under a deep consideration, and the research object must be investigated, not roughly but meticulously and confidentially.

Many persons may consider that S-organisms could emerge with chemical compounds other than [GADV]-amino acids, if the organisms were far simpler and lower than E-organisms. However, I would like to deny the idea here again, because [GADV]-amino acids are as sufficiently as nothing further simple compounds and it would be quite difficult and probably impossible to find out other chemical compounds than [GADV]-amino acids, which can lead to the emergence of life. Therefore I would like to propose that astrobiologists should first search for an indication of α-amino acids, especially [GADV]-amino acids, irrespective of L- or D-amino acids, because every newly born S-life should use proteins composed of α-[GADV]-amino acids, as described in this chapter.

5 Discussion

Many kinds of organisms can be observed in extremely severe environments on the present Earth, as at extremely high or low temperature, in highly acidic or basic water, in water containing an extremely high concentration of salt, in deep sea under a high pressure, and so on. So, many researchers might consider that S-organisms could emerge in such extreme environments in space. Of course, the possibility cannot be denied that S-life, which emerged in such an extremely severe environment similar to an extreme one on the Earth, arrived at the Earth once, looked for a field on the Earth, where the S-life can inhabit, and settled down there.

However, many organic compounds, especially polymers as nucleic acids and proteins, are generally unstable in such an extreme environment, and it would be difficult to retain stably three-dimensional structure in the environment. Such polymers might be also degraded into monomers upon scission of chemical bonds under the environment. Therefore it must be unreasonable to expect that the first S-life emerged in such a severe environment. This also suggests that E-organisms inhabiting under various extreme environments on Earth first emerged in rather an ordinary environment on the primitive Earth, evolved to inhabit in various kinds of extreme environments, and finally adapted to be able to live in the respective extreme environments.

The facts could be given as an evidence that various kinds of special mechanisms are usually equipped to protect organisms living in such extreme environments on Earth. If the extremophiles emerged under such an extremely severe condition in space and inhabited on the primitive Earth, the special mechanism should be observed for organisms living not in an extreme environment but in ordinary environment to adapt to and to protect from the newly encountered mild environment.

It can be concluded that the facts would evidence that some of E-organisms emerged in ordinary environment and adapted to various extremely severe environments in order to live as an extremophile in such a respective extreme environment on Earth. If not the case, it must be considered that the respective S-organisms, which emerged in an extremely severe environment on the respective planets or satellites in space, had arrived at Earth from other planets or satellites having the, respectively, different extreme environments many times and adapted well to the respective environments on the Earth. However, it is well known that all extremophiles on Earth are living under the same life system, that is, which is composed of gene (DNA), the universal or standard genetic code, and protein with 20 natural amino acids except for microorganisms exceptionally using 21st and 22nd amino acids, selenocysteine (Cone et al., 1976) and pyrrolysine (Srinivasan et al., 2002). The fact also indicates that all the extremophiles on Earth are derived from the same ancestor, which emerged under ordinary conditions, because two sets of amino acids used in E- and S-organisms should become different during the respective evolutionary processes at a high probability, although the first amino acids used in the first ancestors of both E- and S-lives would be [GADV]-amino acids.

Taking the discussion in this article into consideration, it is concluded that, if life could emerge anywhere in space, the life always emerges in water environment and with [GADV]-amino acids similarly as E-life emerged on the primitive Earth. The conclusion may not be interesting for many astrobiologists, paleontologists, biochemists, and so on, who are pursuing the riddle on the origin of S-life. However, science should be always to aim at approaching a truth, step by step. The results, at which it arrives with serious experiments and reasonable consideration, are only sometimes interesting and amazing. Generally, fun in scientific activities is felt when approaching to a truth.

Acknowledgment

I am grateful to Dr. Tadashi Oishi (G&L Kyosei Institute, Emeritus professor of Nara Women's University) for the encouragement throughout my research on GADV hypothesis on the origin of life.

References

Barge, L.M., White, L.M., 2017. Experimentally testing hydrothermal vent origin of life on enceladus and other icy/ocean worlds. Astrobiology 17, 820–833. https://doi.org/10.1089/ast.2016.1633 (Epub).

Chandru, K., Obayashi, Y., Kaneko, T., Kobayashi, K., 2013. Formation of Amino Acid Condensates Partly Having Peptide Bonds in a Simulated Submarine Hydrothermal Environment. Vol. 41. Viva Origino, pp. 24–28.

Cone, J.E., Del Rio, R.M., Davis, J.N., Stadtman, T.C., 1976. Chemical characterization of the selenoprotein component of clostridial glycine reductase: identification of selenocysteine as the organoselenium moiety. Proc. Natl. Acad. Sci. U. S. A. 73, 2659–2663.

De Farias, S.T., Rego, T.G., Jose, M.V., 2016. tRNA core hypothesis for transition from RNA world to the ribonucleoprotein world. Life(Basel). 6. https://doi.org/10.3390/life6020015. pii: E15.

Gargaud, M., Martin, H., Lopez-Garcia, P., Montmrle, T., Pascal, R., 2012. Young Sun, Early Earth and the Origins of Life. Springer-Verlag, Berlin, Heidelberg.

Gilbert, W., 1986. The RNA world. Nature 319, 618.

Holm, N.G., Andersson, E., 2005. Hydrothermal simulation experiments as a tool for studies of the origin of life on Earth and other terrestrial planets: a review. Astrobiology 5, 444–460.

Ikehara, K., 2002. Origins of gene, genetic code, protein and life: comprehensive view of life system from a GNC-SNS primitive genetic code hypothesis. J. Biosci. 27, 165–186.

Ikehara, K., 2005. Possible steps to the emergence of life: the [GADV]-protein world hypothesis. Chem. Rec. 5, 107–118.

Ikehara, K., 2009. Pseudo-replication of [GADV]-proteins and origin of life. Int. J. Mol. Sci. 10, 1527–1537.

Ikehara, K., 2014. Protein ordered sequences are formed by random joining of amino acids in protein 0th-order structure, followed by evolutionary process. Orig. Life Evol. Biosph. 44, 279–281.

Ikehara, K., 2016a. GADV Hypothesis on the Origin of Life-Life Emerged in This Way!? LAMBERT Academic Publishing, Saarbrucken, Germany.

Ikehara, K., 2016b. Evolutionary steps in the emergence of life deduced from the bottom-up approach and GADV hypothesis (top-down approach). Life (Basel) 6, 6.

Ikehara, 2017. Life emerged from [GADV]-protein world, but not from RNA world!? Preprints. https://doi.org/10.20944/preprints201712.0170.v1.

Ikehara, K., Omori, Y., Arai, R., Hirose, A., 2002. A novel theory on the origin of the genetic code. A GNC-SNS hypothesis. J. Mol. Evol. 54, 530–538.

Imai, E., Honda, H., Hatori, K., Brack, A., Matsuno, K., 1999. Elongation of oligopeptides in a simulated submarine hydrothermal system. Science 283, 831–833.

Maury, C.P., 2015. Origin of life. Primordial genetics: information transfer in a pre-RNA world based on self-replicating beta-sheet amyloid conformers. J. Theor. Biol. 382, 292–297.

Miller, S.L., Orgel, L.E., 1974. The Origins of Life on the Earth. Prentice-Hall, Englewood Cliffs, NJ.

Raulin, F., Brasse, C., Poch, O., Coll, P., 2012. Prebiotic-like chemistry on Titan. Chem. Soc. Rev. 41, 5380–5393. https://doi.org/10.1039/c2cs35014a. Epub.

Sharov, A.A., 2016. Coenzyme world model of the origin of life. Biosystems 144, 8–17.

Srinivasan, G., James, C.M., Krzycki, J.A., 2002. Pyrrolysine encoded by UAG in Archaea: charging of a UAG-decoding specialized tRNA. Science 296, 1459–1462.

Stryer, L., 1988. Biochemistry, third ed. W.H. Freeman and Company, New York, NY.

Index

Note: Page numbers followed by *f* indicate figures and *t* indicate tables.